BIODIVERSITY: THE ABUNDANCE OF LIFE

Jenny Chapman and *Michael Roberts*

CAMBRIDGE
UNIVERSITY PRESS

PUBLISHED BY THE PRESS SYNDICATE OF THE UNIVERSITY OF CAMBRIDGE
The Pitt Building, Trumpington Street, Cambridge CB2 1RP, United Kingdom

CAMBRIDGE UNIVERSITY PRESS
The Edinburgh Building, Cambridge CB2 2RU, United Kingdom
40 West 20th Street, New York, NY 10011-4211, USA
10 Stamford Road, Oakleigh, Melbourne 3166, Australia

First published 1997

Printed in the United Kingdom at the University Press, Cambridge

Typeset in Palatino 9.5pt

A catalogue record for this book is available from the British Library

ISBN 0 521 57794 2 paperback

Text illustrations by Helen Humphreys

Cover design by Chris McLeod

Cover photo: Box Jellyfish (*Chirondex fleckeri*), North Queensland's most venomous marine organism. Gary Bell, Planet Earth Pictures.

The publisher would like to thank the following for permission to reproduce photographs: Fig 2.3 (above), Biology Media/Science Photo Library; fig 2.3 (below), Dr David J Patterson/Science Photo Library; fig 2.6, Jane Gifford/Natural History Photo Library; fig 2.8 (2), David Scharf/Science Photo Library; fig 3.2, Sidney Moulds/Science Photo Library; fig 3.1 (above) Adam Hart/Science Photo Library; fig 3.1 (below), Dr Morley Read/Science Photo Library; fig 4.3, Anthony Bannister/Science Photo Library; fig 4.6, David Scharf/Science Photo Library; fig 4.11, Ken Lucas/Planet Earth Pictures; fig 4.12, Paul Kay/Natural Science Photos; fig 5.10, Anthony Bannister/Natural History Photo Library; fig 5.11, C Farneti Foster/Natural History Photo Library; fig 6.5, © Tom McHugh/Oxford Scientific Films; fig 6.10 (above), Stephen Dalton/Oxford Scientific Films; fig 6.10 (below left), © Jeff Foott/BBC Natural History Unit Photo Library; fig 6.10 (below right), © Tom McHugh/Oxford Scientific Films; fig 7.3 © M W Powles/Natural Science Photos.

Contents

Introduction

Biodiversity is for A or AS level students doing any of the main syllabus options on the Diversity of Life or Biodiversity. It is also suitable for those learning about microorganisms or plant and animal physiology and as a support text for conservation topics.

The book uses the five kingdom classification of organisms into bacteria, protoctists (including algae), fungi, plants and animals. The main features of organisms from each kingdom are described and illustrated. Explanations of the functions of characteristic structures, as dictated by syllabus requirements, are interspersed with more informal information about the groups.

Sections are included on the definition and recognition of species, past and present extinction rates, and issues of conservation including a case study of the African elephant.

Boxes are used throughout for deeper treatment of various aspects of biodiversity. These include the use of a species identification key, lichens, parasitism, adaptations to life on land and Gaia.

The diversity of life

1.1   What we mean by biodiversity

The word **biodiversity** is shorthand for biological diversity. Our planet is teeming with life and we say that an area has a high biodiversity if it has an abundance of different organisms. One measure of biodiversity is the **number of species** in an area, but the range of **different life forms** (plants, mammals, bacteria, fungi and so on) is also important.

There is concern about threats to biodiversity at both local and global levels. The main threats are from the ever increasing effects of humans on natural habitats and the environment. The consequences of loss of biodiversity may be serious for all life, including humans. To understand the threats to biodiversity we need to understand how biodiversity may change in the future, the risks to species and what is likely to happen as environmental changes occur. This requires us to monitor the earth's biodiversity, how it has been changing through the history of the earth, what it is now and what may jeopardise it in the future.

First of all we must be able to describe it. The basic unit for cataloguing living things is the species. Identifying a species, recording how abundant it is and the **variation** within it will provide a measure of how susceptible the species might be to changes in the environment. A further aspect of biodiversity is the **ecology of the ecosystem** in which each species lives and the inter-relationships with other organisms. This is a huge subject in its own right.

In this chapter we consider species and how they are organised into larger groups. In chapters 2 to 6 these groups are investigated in more detail to give you a better understanding of the diversity of life forms which go to make up the biodiversity of the earth. In chapter 7 the causes of species extinction are investigated, as well as how and why biodiversity should be maintained.

1.2   Identifying species

What is a species?

Species are groups of actually (or potentially) interbreeding natural populations
which are reproductively isolated from other such groups.   E. Mayr, 1942.

This is often given as the classic definition of a species. It is the definition of a
zoologist. Had Mayr been a botanist or a microbiologist, he may have put it
differently. Today we can think of species in terms of genes and gene flow.
Interbreeding individuals have the potential to combine half their genes
with another individual in the population and pass the new combination to
their offspring. Reproductive isolation comes from some barrier which
prevents members of one population recombining genes with members of a
different population.

The concept of species seems simple when animals such as mammals
or other vertebrates are considered. It seems obvious that a mouse is
different from a squirrel. A mouse is also different (but not as different) from
a rat. More similar, but still fairly easy to distinguish, are house mice and
field mice. Mayr's definition enables us to say that house mice and field
mice are different species because they do not interbreed.

Lions and tigers are also different species, but they can interbreed and
produce cubs. However, although lions and tigers can mate, their offspring
are **sterile** so they cannot pass on their genes. In a natural situation this
would be a disaster to the parents as both would have wasted resources
producing an animal that could not continue their line. In nature, lions and
tigers now live in different regions so this problem does not arise. However,
lions once had a much wider distribution including India (there is still one
tiny group left in India). Imagine a situation where lions and tigers lived in
the same area. Any animals which cross-mated would have sterile offspring
and their genes would not be perpetuated. Animals with a tendency to
choose a mate from their own species would mate successfully and their
genes would be passed on to successive generations. Features that become
stronger to prevent cross-mating and create isolation barriers between the
species might include both behavioural and habitat differences.

Most organisms can be classified into distinct groups on the basis of
their appearance (**morphology**), internal processes (**physiology**) and ecological preferences. To prove that they are separate species as defined by Mayr,
the biologist would have to demonstrate that the groups were not interbreeding. In the real world this is hardly ever possible. For example, workers
gathering beetles in the canopy of a tropical forest find many which look
different from anything described earlier. Collecting these beetles usually
involves killing them, so any study of their breeding habits is impossible.
Thus many newly found species are defined on morphological characters
alone; knowledge of their behaviour and breeding habits comes later if time

and money are available. For most species, interbreeding patterns between populations remain unstudied.

Some plant 'species' only reproduce asexually, so individuals do not interbreed at all. In other plants, different 'species' can, and do, cross-breed occasionally and genes from one species may thus be introduced into a population of another species. These examples show that Mayr's rigorous definition of a species may not always fit. For single-celled organisms, Mayr's definition may be even less valid (see chapter 2). However, this does not make the concept of species entirely invalid. If we abandoned it altogether the biologist's job would become impossible. It works most of the time and as long as we are aware that the living world is hugely complex and breaks our rules more often than we would like, then we can proceed with our attempt to categorise the biodiversity around us into species.

Mechanisms of speciation

Understanding how species arise can give us a deeper understanding of the species concept. The classic concept of speciation is that it is gradual: two populations slowly change, becoming more dissimilar in morphology and behaviour until finally the two populations are unable to interbreed. The easiest way this can happen is if the two populations are separated by distance (e.g. if a large population is divided in two by a disaster such as a volcanic eruption) or if a new population is started in an isolated place (e.g. a few seeds float to an island). Speciation of isolated populations is described as **allopatric speciation**.

How two populations become genetically distinct while close together and still able to interbreed occasionally is harder to understand. There may be some strong selective pressure working against the survival of hybrid offspring. For example, populations of grasses found growing on tips polluted with heavy metal are different from those growing on adjacent fields on normal soil. Originally, the tips would have been bare, but they were gradually colonised by grasses able to grow in the toxic conditions. These grasses have a number of biochemical differences to cope with the heavy metals, but they grow slowly. Hybrids *without* the metal tolerance die as seedlings if they grow on tips, while hybrids *with* metal tolerance grow so slowly on normal soil that they are out-competed by normal grass. Speciation in adjacent populations or from two parts of the same population is called **sympatric speciation**.

Other much more abrupt ways a new species may arise are by **mutations** involving large-scale **alteration of the genome**. For instance, if a hybrid is produced from incompatible species (like the tiger and lion mentioned earlier), the offspring are sterile because they contain two half sets of chromosomes which cannot pair correctly in meiosis. If, however, at an early stage in the life of the embryo, the chromosomes had doubled up within the cell, the offspring would have twice as many chromosomes.

These would be able to pair up and the offspring would be fertile. This does not happen in most animals, probably because the genetic control of development is so fine that such a large mutation would cause cell death. However, it does happen in plants, and new species can arise this way instantaneously.

Another type of mutation which can cause speciation occurs when part of a chromosome breaks and re-attaches to another chromosome. If the organism grows and reproduces, the defective chromosome may be unable to pair with a normal one in the offspring. However, in species such as many plants, where vegetative reproduction is possible, such a mutation can be maintained.

In many microorganisms chromosome pairing does not occur (see chapter 2), so all sorts of mutations and alterations in the genetic material can be passed on to form populations with new characters. This makes the species concept particularly difficult to apply.

Variation within species

Within almost every population, we see **variation** between individuals. Organisms may grow to different sizes, or differ in other features. They may have enzymes which function best at slightly different temperatures. Some variation will be due to the environment and is called **phenotypic variation**. Size is a good example as it often relates to the abundance or quality of food or soil nutrients. Other variation will be due to differences in the hereditary material in the different individuals: this is **genotypic variation**.

Usually in nature, external differences in a population appear slight. But if particular characters are selected for, quite remarkable variation can be obtained. This is most obvious in our domestic animals. For example, we have selectively bred many different body forms of dog so that each can best perform some task such as guarding sheep, killing rats, retrieving dead birds, hunting and so on. Yet all these dogs remain one species. The same applies to breeds of farm animals such as sheep. Some sheep breeds can live on very poor diets, others produce fine or tough wool fibres.

Some populations in nature have lost their natural genotypic variation. Best known is the cheetah which has almost no genetic variation left. This is extremely dangerous as it is the genetic variation in a species which enables some individuals to survive new selection pressures. The cheetah is very vulnerable to diseases and changes in its habitat. It is at risk of extinction. Thus variation within a species is important for the maintenance of biodiversity as it reflects the potential of a species to evolve under changing conditions.

1.3   Classification of life forms

Plants and animals

Most of the familiar organisms around us are plants and animals. The differences between plants and animals seem clear and were once thought to be the most important division. Organisms were put into either the **plant kingdom** or the **animal kingdom**. But problems soon arose: what about fungi, which do not photosynthesise but look rather like plants (which is where they ended up), or single-celled organisms with chlorophyll, which have flagella to move around? In 1866, Ernst Haeckel suggested that all single-celled organisms should be grouped in a kingdom of their own. He called them **protists**. But this was not the end of the classification problems.

Prokaryotes and eukaryotes

When optical and electron microscopes improved in magnifying power and resolution, some biologists realised that the most important division is not between animals and plants, but between organisms with two very different types of cell structure. Some single-celled organisms are small and very simple with few internal structures and no membrane around their genetic material. These organisms are called **prokaryotes** as their most obvious feature is that they do not have a nucleus (*pro* = before, *karyon* = nucleus) (see figures 1.1a and 2.1). Other single-celled organisms are larger and more complex with membrane-bound organelles including a nucleus. These are called **eukaryotes** (*eu* = true, *karyon* = nucleus) (see figure 1.1b). All multicellular organisms, including plants and animals, are eukaryotes.

Prokaryotes were the first cellular life to evolve in the 'Archean soup' 3500 million years ago. How this happened is still unclear, but the fossil record suggests that organised cellular mats (probably of filamentous photosynthesising bacteria) existed about 3000 million years ago. Unsurprisingly, fossils of these early life forms seldom show internal details. Eukaryotes appeared about 1500 million years ago. How did they arise?

The most obvious features of eukaryotic cells are their organelles. The largest of these are the nucleus, containing the genetic information for the cell stored in DNA, mitochondria and, sometimes, chloroplasts. Recent studies show that mitochondria and chloroplasts divide independently of the nucleus. Even more interesting is the fact that mitochondria and chloroplasts contain their own DNA and small 70S ribosomes like those found in prokaryotes (eukaryotic cytoplasm has larger 80S ribosomes). It is as if mitochondria and chloroplasts are independent 'organisms' within the cell cytoplasm.

The American biologist Lynn Margulis was one of the first to propose that eukaryotic cells originated from a **symbiotic relationship** of collections of prokaryotes living within another cell. The early prokaryotes may have

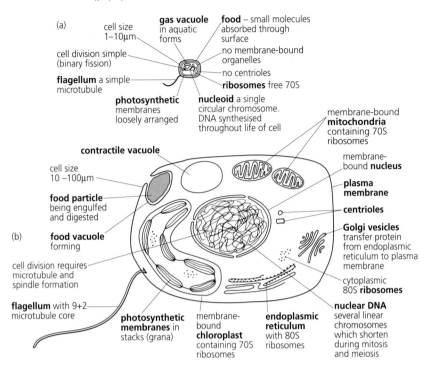

(a)

cell size
1–10µm

gas vacuole
in aquatic
forms

food – small molecules
absorbed through
surface

cell division simple
(binary fission)

no membrane-bound
organelles

no centrioles

flagellum a simple
microtubule

ribosomes free 70S

photosynthetic
membranes
loosely arranged

nucleoid a single
circular chromosome.
DNA synthesised
throughout life of cell

membrane-bound
mitochondria
containing 70S
ribosomes

contractile vacuole

membrane-
bound nucleus

cell size
10 –100µm

plasma
membrane

food particle
being engulfed
and digested

centrioles

(b)

food vacuole
forming

Golgi vesicles
transfer protein
from endoplasmic
reticulum to plasma
membrane

cell division requires
microtubule and
spindle formation

cytoplasmic
80S ribosomes

flagellum with 9+2
microtubule core

photosynthetic
membranes in
stacks (grana)

membrane-
bound
chloroplast
containing 70S
ribosomes

endoplasmic
reticulum
with 80S
ribosomes

nuclear DNA
several linear
chromosomes
which shorten
during mitosis
and meiosis

Figure 1.1 Comparison of (*a*) a prokaryote cell and (*b*) a eukaryote cell, showing the main differences. Figure 2.1 shows a prokaryote bacterial cell in more detail.

started to live symbiotically (e.g. photosynthesising symbionts within a larger host). Today these internal symbionts are so firmly part of the functioning of eukaryotic cells that we can barely recognise that they were once independent. These organelles are called **hereditary endosymbionts**.

The five kingdoms

Appreciating the differences between prokaryotes and eukaryotes led to a re-assessment of the classification of organisms. **Five kingdoms** were proposed by the American ecologist Robert Whittaker in 1969. The prokaryotes became one kingdom and all single-celled eukaryotes were classified as protists; the fungi got their own kingdom; the multicellular plants and animals remained in their original separate kingdoms.

Since Whittaker's suggestions, the picture has changed again. As multicellular algae were considered closer to single-celled algae than to land plants, they were put with the protists into a larger kingdom renamed the **Protoctista**. Then two very different types of Prokaryota were distinguished, so the prokaryotes were divided into the **Archaebacteria** (archaea) and

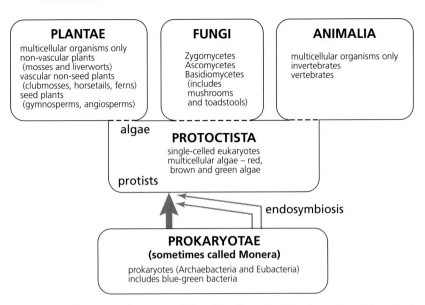

Figure 1.2 The currently accepted five kingdom classification set out to reflect possible evolutionary relationships.

Eubacteria (true bacteria) kingdoms. The justification for this split is that the archaebacteria are as different from the eubacteria as they are from eukaryotes (see chapter 2 for more on these differences).

Today, the most widely accepted, rather conservative, classification is that shown in figure 1.2. The algae which bridge the gap between protists and plants join the protists in the Protoctista, while the archaebacteria and eubacteria are also lumped together. People who are not microbiologists tend to consider large organisms to be more important than microorganisms, and this is reflected in this classification.

Hierarchical classification

So far we have looked at two levels of classification: species and kingdoms. However, the human desire for order has produced a number of intermediate taxa. Thus similar species form a genus (plural genera), similar genera form a family, and so on. The full hierarchy is species, genus, family, order, class, phylum (usually called a division in plants) and kingdom.

Much thought has been given to what these levels mean. The species, as we have seen, may be a fairly 'natural' grouping. The other levels of classification are our way of identifying the evolutionary or **phylogenetic** connections between species. Among the eukaryotes, whole collections of species have evolved by radiative evolution from single early ancestors and our classification tries to reflect this.

We do not have a complete fossil record which we could use to reveal the connections between species, so classifications and relationships are proposed by looking at the characters of present-day living taxa. The trouble is that taxonomists cannot agree on which characters are most important to indicate relatedness, resulting in different versions of classification. Some taxonomists consider all possible characters in the species they are studying. Computers are used to analyse the data and produce the most likely pattern of relatedness of all the taxa in the database. This process is called **cladistics** and is popular with taxonomists at the moment.

Another recent way to study relatedness between organisms is to compare the pattern of base pairs in their DNA, or to look at the final products of that DNA – the amino acid sequences in particular proteins. The percentage similarity between taxa can indicate the degree of relatedness, and the number of differences can even be used to suggest how long ago different evolutionary lines diverged (the **DNA clock** or the **protein clock**).

Box 1.1   Just how many kingdoms are there?

Ever since the classification of living things into five kingdoms in 1969, the flood gates have opened and several different classifications have appeared. Those who create classifications are called **taxonomists**. A **taxon** (plural **taxa**) is the general term for any group of organisms in a classification system. A species is a taxon; so is a genus and a kingdom.

Taxonomists can be classified as 'lumpers' (who consider that **similarities** between groups are most important and therefore put everything into a few large taxa) and 'splitters' (who consider the **differences** between groups to be more important and so create numerous small taxa).

Serious lumpers put all the eukaryotes together and suggest there are only three kingdoms: Archaebacteria, Eubacteria and Eukaryota. At the other extreme, some splitters suggest that there may be eight kingdoms – or even more! For example, they split off a small group of protists which are half way between bacteria and eukaryotes. These Archezoa have a nucleus but lack many organelles including mitochondria; they also have 70S ribosomes in common with prokaryotes. It seems that these protists show the early stages of endosymbiosis similar to the first eukaryotes. Another group of protists get a kingdom of their own (the Chromista) because they have unusual chloroplasts which are surrounded by four membranes instead of the normal two. Chromista are believed to represent a separate event in chloroplast evolution – **secondary endosymbiosis** of one algal cell within another – and thus have a different origin from the chloroplasts in other eukaryotes.

In fact almost everyone who works on obscure groups of prokaryotes and protists thinks their group is so different from everyone else's that they want to create a new kingdom for it!

For example, looking at amino acid sequences in enzymes suggests that the eukaryotes arose about 2000 million years ago, while plants, animals and fungi diverged from each other 1000 million years ago.

Naming organisms

Most organisms are named by using their genus and species names. This avoids confusion as a species may have different common names in different countries, or no common name at all. This two-name system was promoted by Carl Linné, a Swedish naturalist, in the eighteenth century and is called the **Linnaean binomial system**. As the system uses Latin names they are always written in *italics* if placed amongst normal script, *or in* normal type *if within italic text*. When hand-written the name is underlined. The genus is always written with a capital letter and the species name with a small letter. Hence our Latin binomial is *Homo sapiens*. If a binomial is used frequently in a text then it can be abbreviated, after the first appearance, by reducing the genus to its first letter followed by a full stop, for example *H. sapiens*.

When a new species is discovered, it is given a Linnaean binomial and is described, usually in an academic journal. Often a short description in Latin is provided so that even if the journal is in Russian or Chinese, any taxonomist can read it. The binomial name must be unique rather than already belonging to another taxon. Often the species name is descriptive such as *hirsuta* (= hairy) or *aquaticus* (= lives in water). Sometimes it is named after a person. For example, a palaeobotanist in the early 1990s named several fossil plant species after famous ballet dancers, including *Williamsonia margotiana* in honour of Dame Margot Fonteyn. In this example the genus, created in 1870, was also named after a person. There are very strict codes of nomenclature giving instructions on how to name species. Oddly enough, the animal, plant and bacterial kingdoms each have their own independent sets of rules. This means that it is possible for a plant and an animal to be given the same Latin binomial! Before the protists were separated off, it was also possible for very similar organisms to be separated into the plant and animal kingdoms, within totally different hierarchies of classification.

Identification of species

Many more species have yet to be described than are already known. Most are in the highly diverse, but little-studied tropical forests, and most are probably bacteria and insects (see chapter 7). Where plants (flora) and animals (fauna) are well known, books are produced to help us identify organisms. Usually these contain descriptions and drawings of each species plus **identification keys**. These keys are like questionnaires. Each question offers a pair of alternative statements. Once you have decided which one fits your organism, the key directs you to another pair of statements and so on until the specimen has been identified.

Box 1.2   An identification key for woodlice

Here is a simplified key for identification of six of the most common UK species of woodlice. This key only identifies the animals you are most likely to find in your garden. If you become interested in woodlice, a fuller key may be found in *A key to the woodlice of Britain and Ireland* (1991) by Stephen Hopkin available from the Field Studies Council. There are actually 37 species of woodlice in Britain; many are rare or live only in coastal habitats. Most species can be bred in captivity, so if you get really enthusiastic, you can keep them for a while. Place them in a clear sandwich box partly filled with soil, leaf litter and old bark. Add a few small strips of carrot each week and keep the soil moist (but not too wet) with a spray.

How to use the key

When you have gently captured a woodlouse, start with Question 1. If the animal fits the first description, go to question 2; if it fits the second description, go to question 5, and so on. You will probably need a handlens for some questions unless you have very good eyesight. If the animal you have captured does not seem to fit the key at all, then you may have picked up a rare species not covered in this key, so let it go and try another one.

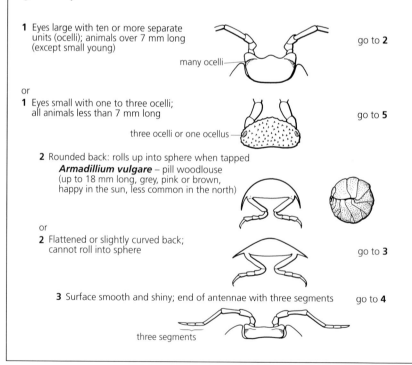

1 Eyes large with ten or more separate units (ocelli); animals over 7 mm long (except small young)

many ocelli

go to **2**

or

1 Eyes small with one to three ocelli; all animals less than 7 mm long

three ocelli or one ocellus

go to **5**

2 Rounded back: rolls up into sphere when tapped
Armadillium vulgare – pill woodlouse (up to 18 mm long, grey, pink or brown, happy in the sun, less common in the north)

or

2 Flattened or slightly curved back; cannot roll into sphere

go to **3**

3 Surface smooth and shiny; end of antennae with three segments   go to **4**

three segments

or

3 Surface rough, end of antennae with two segments
Porcellio scaber – rough woodlouse
(up to 17 mm long, usually slate grey, does not run
immediately when discovered, gardens and compost heaps,
often enters houses)

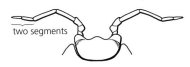

two segments

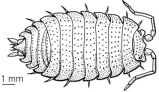

1 mm

4 Smooth curved back with darker central stripe; small (about 11 mm)
Philoscia muscorum – striped woodlouse
(various mottled colours, feels soft when handled,
common in hedgerows and grassland, also occurs
in gardens and woods)

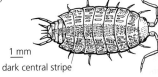

1 mm

dark central stripe

or

4 Flat backed with flange; yellow patches on segments; large (16 mm)
Oniscus asellus – shiny woodlouse
(usually dark greyish brown, does not run immediately
when discovered, under rotting wood and in compost
heaps)

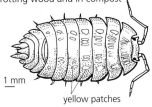

1 mm

yellow patches

5 Eye of three black ocelli; reddish or purplish brown
Trichoniscus pusillus – pigmy woodlouse
(5 mm long, smooth body, damp soil and leaf litter)

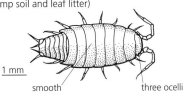

1 mm

smooth   three ocelli

or

5 Eye of one black ocellus; pink to rosy body with double yellow stripe
Androniscus dentiger – rosy woodlouse
(6 mm long, rough body, in garden rubble)

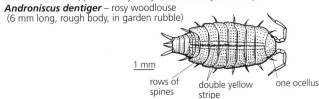

1 mm

rows of   double yellow   one ocellus
spines   stripe

(Key adapted from Hopkin S. 1991 in *Field Studies* **7**: 599–650.)

TWO

Microorganisms

The term **microorganism** is not a precise one and does not occur in any hierarchical classification. It is, however, an extremely useful category, which includes all the organisms that are difficult to study without a powerful microscope. Thus, not only are very small, single-celled organisms called microorganisms, but also some multicellular fungi and algae. Microorganisms include (and therefore this chapter covers) three of the five kingdoms: **Prokaryota**, **Protoctista** and **Fungi**. Although many multicellular algae and fungi are large and therefore not strictly microorganisms, for convenience they are described in this chapter with the rest of the algae and fungi.

There is a huge diversity of microorganisms and many are still to be discovered. We probably only know about a tenth of the world's species of microorganisms. Species are difficult to identify because many microorganisms have not been seen to reproduce sexually. Carl Linné was aware of the existence of tiny prokaryote cells, but he considered them too small to classify and put them in a group he labelled 'Chaos'. Some people who have to study them may wish they had been left there!

Microorganisms are extremely important in our lives, although we do not always know it. Some are harmful – they cause disease in humans, other animals and plants, or are responsible for the rotting of foodstuffs. Many, however, are extremely valuable in aiding digestion, breaking down plant litter, improving soil fertility and, in biotechnology, for synthesising various substances including alcohol and hormones.

2.1 Bacteria (Kingdom Prokaryota)

Classification of bacteria

Bacteria are all prokaryotes. The bacterial cell (see figure 2.1) does not contain membrane-bound organelles such as a nucleus or mitochondria, and the flagella on the surface are simple tubules lacking the 9 + 2 structure of eukaryote flagella.

The bacteria can be divided into two main groups: **Archaebacteria** and **Eubacteria**. They differ in so many ways that many people now think these

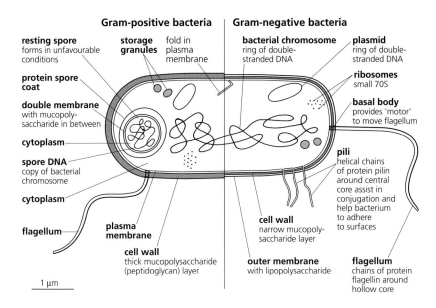

Figure 2.1 The main features of a eubacterium. The diagram shows the two main wall types: Gram-positive on the left and Gram-negative on the right. (Obviously, such walls never occur together like this.) Most internal structures are the same in both types, but only Gram-positive bacteria produce resting spores. Sometimes the whole bacterium is encased in a capsule (not shown on this diagram).

two groups should be classified as separate kingdoms. For example, archae-bacteria have a glycerol-hydrocarbon plasma membrane, eubacteria a phos-pholipid plasma membrane; they also differ in the structure of their transfer RNAs and ribosomal RNA. Here, though, they are put together in the Prokaryota.

The diversity of bacteria is enormous. Some are spherical, others rod-shaped, and all can reproduce by simple division (**binary fission**). Sometimes, elongated cells remain joined after binary fission, producing fila-mentous forms. Certain eubacteria stain purple when a particular stain is used and are described as **Gram-positive**; other similar-looking forms do not hold this stain and are called **Gram-negative**. The Gram-positive bacteria have a thick outer wall of mucopolysaccharide. Gram-negative bacteria have a narrow mucopolysaccharide wall, but this has a lipid and carbohydrate covering with lipopolysaccharides (see figure 2.1) that prevents staining. Although not the plasma membrane, this outer covering has a membranous structure. Archaebacteria have sugar and protein walls and lack the mucopolysaccharide layer.

Some bacteria can form **resting spores** if conditions are unfavourable (see figure 2.1). The cell produces the spore, then dies and breaks open, leaving the spore, which in some species may remain dormant for hundreds of years. There is no change in the genetic component from bacterium to spore, but the spore can survive in conditions which would kill the free form.

Nutrition

Bacteria are abundant everywhere: they occur in the soil, in living organisms, in sea water and in the air. Most bacteria are **heterotrophs**: they need to ingest organic compounds as a source of carbon and energy. Many bacteria (called **saprobionts**) break down organic matter in leaf-litter or carcasses, helping natural debris to decay; other less welcome bacteria cause rotting and build-up of poisonous toxins in our food. Others are important in cheese and vinegar production and some produce substances, intended to kill rival bacteria, which we can use as antibiotics. We need antibiotics because several heterotrophic bacteria are harmful to humans and domestic animals. They invade body tissues and produce harmful toxins or cause cell death. The harmful bacteria (**pathogens**) cause diseases including cholera, tuberculosis and bacterial meningitis.

Some bacteria do not need an external source of organic food: they are **autotrophs** and obtain carbon from carbon dioxide. **Photoautotrophs** can photosynthesise (see below) whereas **chemoautotrophs** obtain energy by breaking down inorganic compounds containing nitrogen, sulphur or iron.

Some bacteria make nitrogen available to green plants. Many such **nitrogen-fixing bacteria** live free in the soil: *Azobacter* and *Rhizobium* fix airborne nitrogen; *Nitrosomonas* converts ammonium to nitrite; *Nitrobacter* converts nitrite to nitrate.

Within the soil, *Rhizobium* thrives best near roots of **legumes**, members of the pea and bean family (Fabaceae). *Rhizobium* is attracted to the very 'leaky' roots by root exudates in the soil. It then invades a root causing the host tissue to multiply to form a **root nodule**. Within the nodule, large numbers of bacteria develop and lose their protective cell walls, forming close contacts with root cell membranes. The bacteria provide nitrogen compounds to the plant. In return the plant provides an enclosed, protected area with a low oxygen content. This is necessary because the bacterial enzyme nitrogenase is denatured by oxygen. The oxygen is removed by a molecule made by the plant, similar to blood haemoglobin and called **leghaemoglobin**. The relationship between the legume and its nitrogen-fixing bacteria is mutually beneficial. This is a type of symbiosis called **mutualism** .

Using legumes as green manure (ploughing in the plants while they are still growing) has been a farming practice since Greek and Roman times. When the plants die and decay, they release the nitrogen from the nitrogen-fixing bacteria. Geneticists are trying, through genetic engineering, to introduce nitrogen-fixing capabilities into crops such as wheat. However, the process is governed by several genes and no eukaryotic cell has yet been found with nitrogen-fixing abilities – so geneticists may be faced with a difficult task.

Photosynthetic bacteria

Several bacteria contain pigments that react to light. The best known are **cyanobacteria**, sometimes wrongly called blue-green algae (algae are eukaryotes). Cyanobacteria are found in many habitats including sea water, the surfaces of wood and soil, just under the surface of rocks in the Antarctic at subzero temperatures, and in hot freshwater springs at 70°C! The cells contain free membranes, not arranged in a chloroplast, containing chlorophyll a and other pigments which give a bluish-green colour. Cyanobacteria do not have flagella; they move by gliding. They may be related to Gram-negative eubacteria.

Cyanobacteria and other photosynthetic bacteria (**green sulphur bacteria** and **purple sulphur bacteria**) use carbon dioxide as a source of carbon. Cyanobacteria photosynthesise in a similar way to green land plants: they split water to provide a reducing environment for the carbon and release oxygen. Sulphur bacteria contain a different pigment, named **bacteriochlorophyll**, and they use sulphur or hydrogen gas to reduce carbon dioxide and release sulphuric acid or water. They require an **anaerobic** (oxygen free) environment. Because these bacteria cannot grow in the presence of oxygen they are called **obligate anaerobes**. The bacteria cannot get rid of oxygen, which builds up in the cell and rapidly becomes toxic. Some obligate anaerobes survive high oxygen conditions by forming a resting spore (see figure 2.1).

Some pigmented bacteria (**purple non-sulphur** and **green non-sulphur bacteria**) use light as an energy source, but require organic acids or alcohols as a carbon source. The green non-sulphur bacteria are a small group with only one genus. They have chlorophyll a and b and carotenoids on their membranes. These pigments are closer to those of eukaryotes than to the pigments of other bacteria.

One remarkable group of archaebacteria, the **halobacteria**, live in extremely salty environments such as inland salt lakes. Their plasma membranes contain a purple pigment called **bacteriorhodopsin** (similar to rhodopsin in human retinal cells) which appears to function like chlorophyll. Normally, halobacteria use oxygen in the electron transport chain (oxidative phosphorylation), but the light reaction is a back-up system when oxygen is in low supply. The presence of rhodopsin-like substances raises interesting questions about endosymbiosis and the evolution of the primitive eye.

Hot bacteria

Some of the most remarkable bacteria live in extremely hostile environments: in hot springs and deep sea vents where water is heated, sometimes to boiling point, by the closeness of molten magma in the earth's crust. These waters are also very rich in dissolved minerals and, depending on the rock the waters filter through, they can be very acidic (pH 2) or very alkaline (pH 9). Both eubacteria and archaebacteria are found in these conditions and some of the archaebacteria thrive in temperatures above 80 °C. The bacteria have enzymes and cell structures which function at much higher temperatures than other bacteria.

Such heat lovers (**thermophiles**) may one day be useful for producing biotechnological products at high temperature. Possible uses include the production of ethanol from polysaccharides, sewage treatment (where high temperatures will aid sterilisation by killing many other microorganisms), composting waste and extracting metals from low-grade ores. Thermophilic bacteria are already used in the polymerase chain reaction to multiply up minute amounts of DNA for DNA fingerprinting.

Gut bacteria

There is an enormous and varied microbial assemblage within the animal gut. This is especially important for herbivorous mammals such as grass eaters (e.g. sheep and deer) and broad-leaf eaters (e.g. colobus monkeys). Gut bacteria and protists (see page 19) possess cellulase enzymes which digest cellulose from plant cell walls. This activity makes sugars available to the animal so it can survive on a poor quality diet. This is an example of mutualism – the bacteria have a place to live and constant food supplies; the animal gets the extra sugars.

Escherichia coli is a Gram-negative, flagellated, rod-shaped bacterium. Millions of them live in each human intestine. It is a close relative of the food-poisoning bacterium *Salmonella enteritidis*. Most *E. coli* populations are only occasionally harmful (in very young or old people), when they multiply in the urinary system. Some populations are more aggressive, and these different strains can cause a number of illnesses including gastroenteritis and kidney failure. Normally, *E. coli* may be beneficial to the host as a source of vitamin K. However, if someone has a burst appendix, gut bacteria spill into the body cavity and cause peritonitis.

Salmonella is a different matter altogether. If someone eats food heavily infected with *S. enteritidis*, then the bacterial toxins cause classic food-poisoning symptoms of vomiting and diarrhoea. The bacterium is prevented from multiplying by rapid freezing and is killed by temperatures of 60 °C, so if food is properly treated, not enough bacteria will be present to cause problems. The poisons, which irritate the gut lining, are the lipopolysaccharides in the outer covering of the Gram-negative wall. Because they are in the wall of the bacterium they are called **endotoxins** and are released when the cell dies and is digested in the gut. If enough bacteria are ingested, *S. enteritidis* can be fatal. Many *Salmonella* species cause disease, including *S. typhi* which is responsible for typhus fever.

Another bacterium responsible for food poisoning is *Staphylococcus aureus.* This species *secretes* toxins (thus called **exotoxins**) into the gut. These exotoxins are stable at high temperatures, so they remain in poorly cooked foodstuffs. *S. aureus* produces an enzyme which allows it to invade the body. It can enter through the gut, or via hair bases to form boils, and may infect wounds after surgery.

Bacteria can multiply very rapidly. The bacteria in your gut have already been through the same number of generations as we have gone through since humans evolved! Their amazing rate of cell division provides bacteria with an advantage over longer-lived organisms both in numbers and in the rate at which they can acquire new traits. Many different anti-biotics are used to kill harmful bacteria, but if just a few bacteria amongst the millions are more resistant and survive, they can soon multiply up and cause another infection. Microbiologists are concerned about the over-use of antibiotics in animals and humans. Often, antibiotics are given to healthy farm animals to *prevent* disease. Several bacteria (e.g. *Staphylococcus aureus*) now have strains immune to most antibiotics. This does not bode well for the future.

Plasmids

Plasmids are double strands of DNA found in bacterial cells. They are inde-pendent of the main bacterial chromosome and replicate separately from it. They are passed on during normal cell division and conjugal transfer (see below) and carry genes from one strain to another and even between species. This is important because plasmids often contain genes conferring resistance to antibiotics. They can also make some bacteria more virulent.

Some plasmids are beneficial to us: the genes which cause legumes to produce root nodules are carried in the plasmids of *Rhizobium*. Some plasmids carry resistance to toxic heavy metals such as mercury or lead; others confer the ability to digest large organic molecules. These plasmids may be useful in biotechnology for cleaning up metal contaminations and oil spills.

Reproduction

Despite being so simple, bacteria have extremely complex and varied ways of obtaining and passing on genetic material, often from other strains or even other species. There are four main ways:

1 **Binary fission** is the usual, asexual form of reproduction. The main chromosome and the plasmids are copied and move to opposite ends of the cell, which then divides in two. This is not mitosis as no system of microtubules and centromeres is involved. No recombination of genetic material occurs: the new bacteria are identical to the original bacterium.

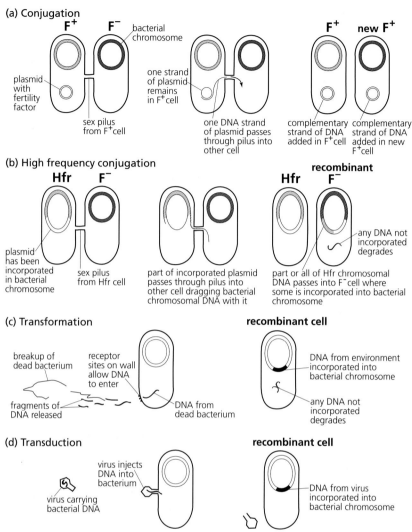

Figure 2.2 The different ways in which a bacterium such as *E. coli* can obtain new genetic material. (*a*) Conjugation between a bacterium with the fertility factor for conjugation (F⁺) on a separate plasmid and a bacterium lacking the factor. (*b*) Conjugation between a bacterium with the plasmid containing the F factor incorporated in the bacterial chromosome, and a bacterium lacking the factor. In this situation there is a high frequency of recombination of genetic material so the donor bacterium is called high frequency (Hfr) donor. Usually not all the chromosome passes to the other bacterium before it breaks, so the recipient may not get all the F⁺ genes. (*c*) Transformation occurs when a living competent bacterium takes in DNA from the environment and incorporates some into its genome. (*d*) Transduction occurs when a virus carrying bacterial genes injects them into a bacterium. The genes will have been trapped inside the viral coat by accident when a virus attacked another cell. Because the genes are not viral, they may be incorporated into the bacterial genome without damaging the bacterium. See figure 2.9b for an illustration of how a virus attacks a bacterial cell.

2 **Conjugation** involves the transfer of genetic material from one bacterium to another and is mediated by plasmids. The donor must have a section either on a plasmid (see figure 2.2a) or set into the main chromosome (see figure 2.2b) called the **fertility** (or **F**) **factor** and the recipient must lack it. Recombination occurs in the recipient, which becomes a donor strain. Conjugation occurs in a number of bacteria including *E. coli*.

3 **Transformation** occurs when the bacterium obtains DNA directly from the environment (see figure 2.2c). DNA is released by the **lysis** (breaking-open) of dying bacteria. The bacterium has to be capable of taking in the DNA, which it recombines with its own genetic material. Such bacteria are described as **competent** and include *Streptococcus*, *Staphylococcus* and *Rhizobium*. *E. coli* can be made competent under laboratory conditions and persuaded to take in genes that enable it to make substances we require.

4 **Transduction** takes place when new genetic material is carried in a virus (see page 32) and is injected into the host cell (see figure 2.2d). The genetic material comes from the last bacterial cell the virus infected and is trapped by accident within the viral coat. Transduction is not under the control of the bacterium but, once inside the cell, the genetic material can be incorporated into the bacterial chromosome. Transduction can be induced in *E. coli* to provide strains which are biochemically useful (see page 35).

2.2   Protists (included in Kingdom Protoctista)

Characteristics of protists

Protists are not really a natural group but are lumped together because they are single-celled (unicellular) eukaryotes. It is, however, a useful group which prevents the splitters from getting the upper hand. Protists are single-celled organisms closely related to fungi, algae and animals. This makes it a very diverse group.

Because of their vulnerability to dehydration, most protists live in moist environments: in the sea or freshwater, moist soils and as parasites within other organisms. Some are heterotrophs, but many have chloroplasts and a variety of photoreactive pigments and so are capable of photosynthesis. Some photosynthesise but also require certain organic compounds (such as vitamins) from the external environment. **Gas exchange** of oxygen and carbon dioxide during respiration (and photosynthesis in some forms) is by **diffusion** through the plasma membrane.

Some photosynthetic protists and unicellular green algae live in **symbiotic relationships** with a variety of organisms including corals, flatworms, flies and fungi (see box 2.1 on page 31). These relationships are not always beneficial to the protists, which get rejected or even digested under certain circumstances.

Many protists live out their lives floating in the sea as **plankton**. The **phytoplankton** (photosynthetic) are the main **primary producers** of the seas, feeding the **zooplankton** (heterotrophic) and, via them, the rest of the food chain. Photosynthetic **dinoflagellates** bloom in vast numbers, when conditions are favourable. There can be so many that the sea looks red from their pigments. These red tides are bioluminescent and glow blue at night if disturbed by waves or the propellers of ships. **Diatoms** also float in the phytoplankton; these are beautiful organisms encased in ornate silica boxes. Another group, **coccolithophorids**, are photosynthetic flagellates which are covered in tiny protective shields of calcium carbonate (known as coccoliths). Although now rare, they were once abundant in the oceans over 100 million years ago. The white cliffs of Dover are made of countless billions of tiny fossil coccoliths. The decayed bodies of huge blooms of phytoplankton, buried and compressed under tons of sediment, are almost certainly the source of our oil and natural gas deposits.

Protozoans

Many protists are motile heterotrophs that lack photosynthetic pigments and feed on organic substances. A handy label for these is **protozoans**. Protozoans are classified into phyla mainly on the basis of their methods of locomotion.

Species in the genus *Amoeba* (Phylum Rhizopoda) are protozoans with very mobile shapes (see figure 2.3a). They move by pushing out projections of the cell called **pseudopodia** into which the rest of the cytoplasm flows. **Rhizopods** eat by pushing out pseudopodia which engulf a particle of food, so forming a food vacuole. The trapped food is digested by enzymes released into the vacuole from lysosomes. The soluble products of digestion are then absorbed. In fresh water, because there is no rigid external wall, water continually enters the cell by osmosis and must be actively expelled via a contractile vacuole. Some rhizopods, called **foraminifera**, secrete a protective shell. Many of these shells are found fossilised in rocks.

As a means of locomotion, pseudopodia are fairly inefficient and slow. Some protozoans, however, can move very rapidly by beating a forest of tiny projections called **cilia**. **Ciliates** (Phylum Ciliophora) include *Paramecium* which, as you can see in figure 2.3b, has a much more organised structure than *Amoeba*, including a 'mouth' (**oral groove**) which collects food particles. Some ciliates lose their motility and live a stationary life held up in the water by a stalk. They use their cilia to waft food particles into a much enlarged oral groove rather like a hollow cup.

Figure 1.1b illustrated a 'typical' eukaryotic cell with a large flagellum of the complex 9 + 2 strand structure. Protozoans with one or more such flagella are called **flagellates** (Phylum Zoomastigina). Flagellates are often

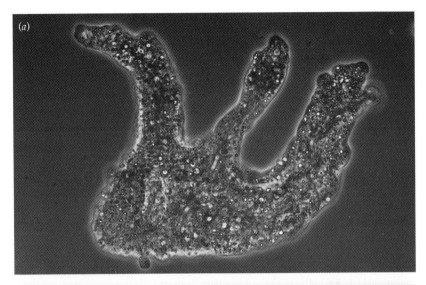

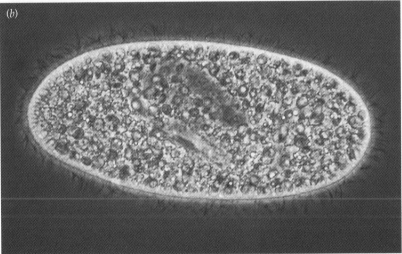

Figure 2.3 (*a*) Light micrograph of *Amoeba* with three long pseudopodia. (*b*) *Paramecium*. The photograph is taken using phase contrast which allows the many cilia around the protozoan to be seen. Note the more regular oval shape compared with *Amoeba*.

much smaller than ciliates and many are parasites. Sleeping sickness, for example, is caused by the flagellate *Trypanosoma*.

The **sporozoans** (Phylum Apicomplexa) are almost all parasites. They have diverse and often complex life cycles and a wide range of forms. In fact, there may be several different forms in the life cycle of just one species. *Plasmodium falciparum* is one sporozoan from a whole genus which causes malaria. The protozoan is carried about in the gut of the mosquito (the **vector**) where male and female forms fuse to form oocysts. These burst to release spindle-shaped sporozoites which move through the mosquito's body to its salivary glands. When the insect bites a human to gather blood, the sporozoites are injected into the human bloodstream, which transports them to the liver where they multiply asexually within the liver cells. Another form – merozoites – are released from the liver cells back into the bloodstream and invade red blood cells. Again the cells multiply, destroying many red blood cells and releasing toxic wastes which produce fever in the infected person. Some merozoites remain in the blood cells and develop into a sexual phase. These transfer to the gut of the mosquito when it takes a meal of human blood, thus starting the cycle again.

Euglenoids (Phylum Euglenophyta)

One group of protists, the euglenoids, were more responsible than any other for raising early doubts about classifying all organisms as plants or animals. Some euglenoids, such as *Euglena* (see figure 2.4) have a light-sensitive **photoreceptor** which can detect the direction of a light source using a light-blocking patch called the **eye spot**. As the cell moves along by whipping a single long flagellum, it spins and the eye spot rotates along with the body. Every time the eye spot gets between the light source and the photoreceptor, it cuts off the light. The cell usually seeks light, but avoids very strong light.

The problem in early classification was that euglenoids have both plant and animal characteristics. They have no cell wall but a stiff layer,

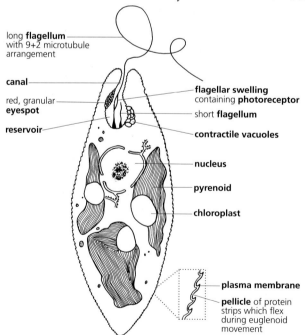

long **flagellum** with 9+2 microtubule arrangement

canal

red, granular **eyespot**

reservoir

flagellar swelling containing **photoreceptor**

short **flagellum**

contractile vacuoles

nucleus

pyrenoid

chloroplast

plasma membrane

pellicle of protein strips which flex during euglenoid movement

Figure 2.4 The internal structure of *Euglena*. This example contains several large chloroplasts; others lack chlorophyll but look similar otherwise.

called the pellicle, just inside the plasma membrane. This gives shape to the cell but also flexibility so that euglenoids can move by wriggling (euglenoid movement) when the flagellum is at rest. They contain a contractile vacuole and storage granules of paramylum (a polysaccharide). Some euglenoids have chloroplasts containing chlorophyll a and b; others lack chlorophyll, yet resemble the photosynthetic forms. If kept in the dark for several generations, *Euglena* lose their chloroplasts and switch nutrition from partially autotrophic to totally heterotrophic.

2.3   Algae (included in Kingdom Protoctista)

The algae are a very diverse group of pigmented photosynthetic organisms. Many are multicellular and some are very large. The multicellular algae have an almost unbelievable variety of extremely complex patterns of reproduction. Most include **alternation of generations** between a haploid **gametophyte** (with one set of chromosomes) and a diploid **sporophyte** (with paired sets of chromosomes). You may be relieved to know that the complexities of most algal reproduction are beyond the scope of this book, but you will find more on alternation of generations in the next chapter (see box 3.2 on page 44).

Algae are classified, on the basis of the colours produced by their pigments, into green, red and brown algae. Many are familiar to us as the slippery seaweeds which grow on rocky shores below the highest tide levels. The green algae, more than any other group, show the stages of evolution which probably occurred in the Palaeozoic over 400 million years ago, before true vascular plants evolved. They range from single-celled and colonial forms to complex multicellular seaweeds. Some green algae have invaded terrestrial habitats although they are not officially classified as 'land plants'.

Green algae (Phylum Chlorophyta)

The green algae are by far the most diverse algal group. They live in sea water, fresh water and moist terrestrial places such as tree bark and wooden fences. Forms range from unicellular, filamentous and colonial to large multicellular seaweeds. The vast majority of green algae live in fresh water. They contain chlorophyll a and b (the same pigments as in vascular plants), and carotenoids, which sometimes give them a reddish appearance.

Chlamydomonas (see figure 2.5) is a single-celled chlorophyte with two flagella which lives in ponds, puddles and melting mountain snow-fields. It has a light-sensitive region and eye spot similar to *Euglena* (see figure 2.4). In fact, at first glance these green algae appear very similar to the euglenoids, but *Chlamydomonas*, like all algae, has a cellulose cell wall (euglenoids have no cell wall), stores starch (not paramylum) and never becomes heterotrophic in the dark.

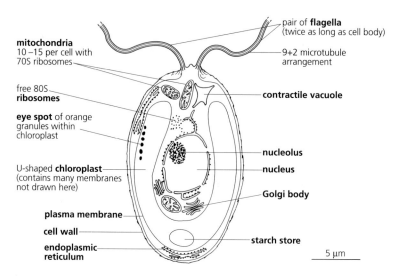

Figure 2.5 The structure of the green, unicellular alga *Chlamydomonas*. (Redrawn from Pickett-Heaps J. 1975 *Green Algae*, Sinaurs Assoc. Inc.)

Another green alga, *Chlorella*, lives in ponds and damp soil, and is very similar to *Chlamydomonas*, except that it lacks flagella and has only one, highly branched mitochondrion. It is famous because part of the biochemistry of photosynthesis (the Calvin cycle) was discovered using this organism. It only reproduces asexually, forming a tough coat inside which a few spores develop, unlike *Chlamydomonas*, which also undergoes sexual reproduction.

Volvox is a colonial organism consisting of hundreds or thousands of *Chlamydomonas*-like flagellates grouped into a hollow sphere. These colonial associations give us a glimpse of possible early steps along the path from unicellular eukaryotes to complex multicellular forms. Another multicellular line may have begun with cells which failed to separate after division resulting in elongated strands or filaments. A living example is *Spirogyra*, whose haploid cells in the strand each carry two spiral chloroplasts.

Stoneworts are important freshwater chlorophytes. They have a stem with whorls of branches and a calcareous substance in the cell walls. When they decay they produce a fine mud which fills the bottom of lakes. The terrestrial green plants are now thought to have evolved from an ancestor of the present-day stoneworts.

Red algae (Phylum Rhodophyta)

Although red algae seem rather rare around Britain's coast, they are extremely abundant seaweeds in tropical and warmer temperate waters. More than half the known seaweed species are rhodophytes. Red algae contain chlorophyll a and d but their red colour comes from the protein **phycoerithrin** and carotenoids. The mix of pigments in red seaweeds allows them to thrive in deeper water than other seaweeds: down to 200 m depth where only blue and green light penetrates. In low light intensities, the amounts of protein pigments increase. The relative amounts of pigments produce a colour range from yellowish-green to purple. Some red algae (e.g. *Corallina*) have calcium carbonate crystals to strengthen the cell walls; this gives them a coral-like feel and appearance. *Corallina* lives in rockpools where wave action is strong.

A small number of rhodophytes live in freshwater streams, and a few single-celled forms occur. Most unusual, though, are **parasitic** forms which live as nodules attached to other seaweeds. They have lost most or all of their chloroplasts and obtain food directly from their host – yet another example of the strong tendency in evolutionary lines to try for a 'free meal' !

Agar, used as a culture medium in biotechnology, comes from red algae. Agar is also used in jelly, mayonnaise, processed cheese and paint.

Brown algae (Phylum Phaeophyta)

The brown algae are mostly restricted to the sea; very few live in fresh water and all phaeophytes are large multicellular structures. They dominate the shore and low-shore zones in cold waters at high latitudes (see figure 2.6).

Figure 2.6 The rocky shore of Loch Torridon, Scotland. The tide is out exposing several species of brown algae.

Many, especially those in brackish water, only reproduce vegetatively. Tropical groups include the huge floating weeds of the Sargasso Sea.

Brown algae possess chlorophyll a and c, plus a carotenoid **fucoxanthin,** which gives them a dark greeny-brown colouring. The largest forms are very robust with a swollen base or **holdfast** for attachment to rocks, a long tough stem of interwoven filamentous cells called a **stype,** and a flattened **lamina** of rounded parenchyma cells. Some brown seaweeds grow over 20 m in length.

Genera common around the British coast include *Laminaria* (kelps) and *Fucus* (bladder-wracks). The ashes of burned kelp have been used for centuries as a glaze for pottery or as a source of iodine, and whole kelps are used as fertiliser and a source of food for farm animals and humans, especially in the Far East. A kelp product, alginate, is used as a stabiliser in ice cream and paints.

2.4   Fungi (Kingdom Fungi)

The fungi we know best are mushrooms – the edible reproductive structures of various species. Some, such as underground truffles, are considered a culinary delicacy and are very expensive to buy. People who study fungi, rather than just eat them, are known as **mycologists.**

Most fungi consist of a branching network, or **mycelium,** of thin filaments called **hyphae.** In some taxa these hyphae are divided by **septa** (they are **septate**); in other taxa the hyphae have no divisions (they are **aseptate**). The walls of the hyphae are made of the polysaccharide **glucan** with a network of **chitin** and proteins. Chitin is a substance more usually associated with arthropod cuticles (see page 73). Within the hyphae, the cytoplasm contains numerous mitochondria, vacuoles, bound and free 80S ribosomes (larger than those in bacteria), abundant endoplasmic reticulum and scattered nuclei. Golgi bodies are usually absent.

Most mycelia look alike, so fungi are identified by their spore-producing structures. Fungi lack photosynthetic pigments, so all have to feed heterotrophically as saprobionts, parasites or mutualistic symbionts. Some parasites, including moulds, kill the host cells. Fungi are widespread in the soil and many are important as symbionts with plant roots (see the section on mycorrhizae on page 32).

Mostly moulds (Phylum Zygomycota)

Moulds are usually saprobionts. They grow on organic matter including woodland litter and dung and prepared foodstuffs, such as cake and bread. Some cause the rotting of ripe fruit and vegetables. Their hyphae are usually aseptate.

Mucor is a mould which lives on dead organic matter including bread and fruit; it digests the substrate by secreting enzymes from the tips of its hyphae. *Mucor* reproduces asexually by sending up a miniature forest of vertical hyphae (see figure 2.7). At the tip of each hypha, a swelling develops within which the nuclei divide several times and each swelling becomes surrounded by a spore wall. The mature spores are released and are spread by insects or air currents. Sexual reproduction can occur if two adjacent mycelia are of different strains (see figure 2.7).

Mildews (Phylum Oomycota)

Oomycetes oscillate between classification in the Protoctista and the Fungi. They differ from most fungi as their aseptate hyphae have scattered *diploid* nuclei and *cellulose* cell walls; they also have flagellated sexual cells (gametes). The fine hyphae sometimes seen radiating from a dead animal floating in a pond are oomycetes (**water moulds**). Other mildews cause plant diseases. In fact one humble mildew changed the course of history. The potato-blight, *Phytophthora infestans*, was responsible for the failure of European potato harvests in the nineteenth century. The extent of the blight in Ireland led to terrible famines and forced many families to emigrate, some to successful lives in America.

Yeasts and more moulds (Phylum Ascomycota)

Ascomycetes include fungi with mycelia which have perforated septa; many have a mould-like form. They usually reproduce sexually by fusion of two of their own haploid nuclei. The diploid nucleus then divides by meiosis and eight new haploid spores develop in a special sac called an **ascus**, at the end of a hypha. Truffles are the underground reproductive bodies of ascomycetes and are filled with scattered asci.

Yeast (*Saccharomyces*) has an unusual structure for a fungus. It consists of rounded cells which bud off to form a small branched system (see figure 2.8a). These cells are diploid (unlike the hyphae of most fungi). Yeasts grow well in oxygen; they can also survive under oxygen-free conditions by respiring anaerobically. They are called **facultative anaerobes** because they switch biochemical pathways to use oxygen if it is present. When respiring anaerobically, yeasts convert sugars into ethanol and carbon dioxide by a process called **fermentation**:

$$C_6H_{12}O_6 \rightarrow 2C_2H_5OH + 2CO_2$$

In brewing, yeast produces the ethanol (alcohol) in beer and wine. The carbon dioxide is of little importance (except in champagne) and is usually allowed to escape. In bread making, the ethanol is just a by-product which is driven off during cooking; the carbon dioxide puts bubbles in the bread during the rising process. Wild yeasts live in the soil and on ripe fruit; the

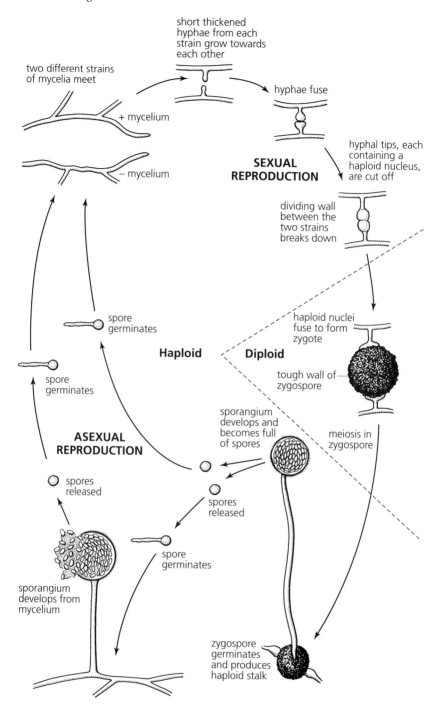

Figure 2.7 The life cycle of *Mucor*. *Mucor* is a zygomycete fungus which produces both sexual and asexual spores. When sporing, the mould has a fuzzy green, bluish or black appearance.

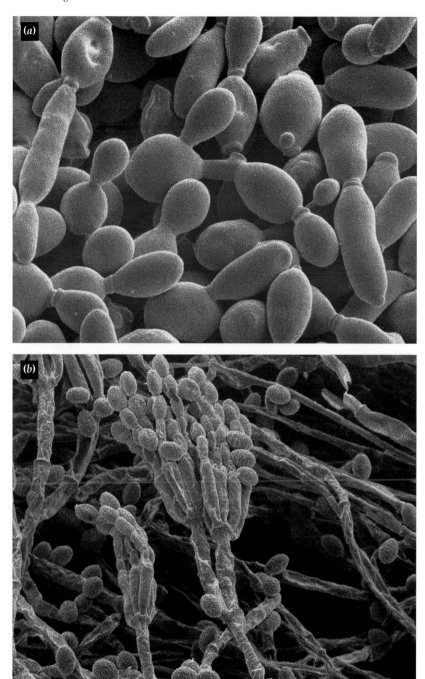

Figure   2.8 (*a*) Scanning electron micrograph of budding yeast (magnification x2500). This species, *Saccharomyces cerevisiae*, is used in brewing and breadmaking. (*b*) Scanning electron micrograph of a conidiophore of the blue mould, *Penicillium* (magnification x1200). The oval conidia are budded off and released from the branch ends. The hyphae of the blue mould can be seen in the background.

'bloom' on a grape or plum is a yeast coating. However, today's commercial yeasts have been specially selected and most bread in Britain is made using genetically engineered yeasts.

The blue mould *Penicillium* is also an ascomycete. It usually reproduces asexually by releasing tiny spores called conidia from branch-like conidiophores (see figure 2.8b). Sexual reproduction is very rare. It produces the antibiotic **penicillin** which weakens the cell walls of growing Gram-positive bacteria, causing the bacteria to burst (lyse). Mature bacteria are not affected. The outer membranous covering of Gram-negative bacteria protects them from penicillin, so most Gram-negative bacteria are not affected by this antibiotic. The antibiotic effect was first noticed by Alexander Fleming in 1928, in a Petri dish where he was growing bacteria. A conidiospore must have entered the dish by accident. Fleming noticed that around the mould was a halo with no bacteria. The resulting discovery saved the lives of many wounded soldiers during the Second World War, when bacterial infection would otherwise have led to gangrene and death. However, only 20 years after its discovery, many bacteria were already showing immunity to this antibiotic (see page 17).

Penicillium moulds are also used to make blue cheeses such as Stilton and Roquefort.

Mushrooms and toadstools (Phylum Basidiomycota)

The wild and cultivated mushrooms we eat are the sporing bodies (basidiocarps) of basidiomycetes. The cap of the common edible mushroom (*Agaricus*) protects the delicate gills underneath from the rain. The gills hold the spore-bearing sacs. A spore print can be made by placing an open flat-capped mushroom, without its stalk, with gills facing downwards on white paper. After a day or two, a pattern of brown spores will appear on the paper. Not all basidiocarps are edible; several woodland species are highly poisonous to humans and wild mushrooms should only be gathered by an expert in the identification of basidiomycetes.

Basidiomycetes have hyphae with septa and usually two nuclei per section. When a new cell is cut off at the end of a growing hypha, both these nuclei divide. The new cell wall separates one pair of nuclei but the other pair remain in the hyphal tip. The hypha grows a little side arm which loops back to the new cell, allowing one from this pair of nuclei to enter the new cell. This leaves a little bridge, called a 'clamp connection,' curving over the septum between the cells. It is this feature which allows hyphae to be identified as basidiomycetes.

Some parasitic groups of basidiomycetes do not form basidiocarps. They cause plant diseases and have the colourful names of **rusts** and **smuts**. Rusts produce orange-red spores and smuts powdery black spores. They can reduce crop yields, especially in monocultures of grain crops.

Box 2.1   Lichens

Lichens, although easily overlooked, are common on undisturbed surfaces such as gravestones and old walls. They can look rather unassuming – like something that has been spilt and dried there – but actually they are a fascinating example of mutualism.

Lichens are formed by a close association between a fungus (usually an ascomycete, sometimes a basidiomycete) and either a cyanobacterium or a green alga. The photosynthesising partner fixes carbon, which provides the fungus with carbohydrates. Nitrogen compounds come from airborne dust, but a small amount may be produced by the cyanobacterium if this is present. It is not obvious exactly what the photosynthetic organism gets from the relationship – perhaps an extension of its habitat range, or maybe it is not benefited at all, but is 'held prisoner' by the fungus.

Lichens grow in various shapes: crusty, finely branched or a curled foliose shape rather like a dried-up oak leaf. A fungal species sometimes forms a different-shaped lichen depending on whether it is with a cyanobacterium or a green alga. The larger lichens often festoon rocks and trees.

Lichens grow remarkably slowly. This is partly because their only source of carbon compounds is the photosynthesising partner, which usually forms only about 10% of the total biomass. Another reason is that lichens inhabit such exposed sites that they often dry out and are scorched in summer or frozen in winter. One way of estimating the age of lichens is to measure their diameter on gravestones and correlate these with the time the stone has been there (obtained from the year of death carved on it). This will provide a measure for other sites. Try dating some lichens by this method, then use it to date a wall or building. What errors might arise with this method and how could they be overcome?

Lichens have many uses. In industrialised nations they are **biological indicators** of pollution, as they are extremely sensitive to atmospheric contamination, especially sulphur dioxide. Delicate foliate lichens die first, then others, so the fewer lichens are found, the more polluted the air. By now, many industrial countries are so polluted that they have lost 25% of their lichens. One of the most tolerant is a dull grey-green encrusting lichen *Lecanora conizaeoides* (lichens are given their own binomial despite consisting of two different species). *Lecanora* was unknown 200 years ago, but is now abundant around cities. It is thought to have grown around sulphur springs, a very rare habitat, until pollution gave it the chance to become widespread.

Lichens are a valuable food source in the Arctic. Despite growing slowly, they are the main food of reindeer, which browse on the large foliate lichens growing on soil and rocks. Quite large herds can be sustained by this insubstantial vegetation. Lichens are also used to provide soft-coloured dyes for cloth such as Harris tweed.

Mycorrhizae

Fungi form two types of mutualistic association with plant roots. Some associations (called **endomycorrhizae**) have hyphae which penetrate root walls; the other group remain extra-cellular (**ectomycorrhizae**). Orchids need endomycorrhizae to survive, even to germinate and grow into seedlings. An orchid seedling relies on the fungus for organic compounds for up to ten years, until its first leaf is functional. Shrubs (e.g. heathers) on poor soils almost always possess ascomycete endomycorrhizae.

One form of endomycorrhizae are the **vesicular-arbuscular mycorrhizae** (VAM for short). The fungal hyphae are zygomycetes which develop special structures within the plant root. Vesicles are swollen hyphae which act as storage places for the fungus. Arbuscules are finely branched little structures which penetrate root-cell walls but not the plasma membrane. They branch out to form a close, many-fingered connection with the plasma membrane, and through this connection the plant obtains phosphates. The host plant gains considerably from this association as the fungal hyphae increase root effectiveness by extending several centimetres into the soil. Nutrient uptake is improved, especially phosphate supply, water stress is reduced, and there may be some additional nitrogen availability due to nitrogen fixation. There is also evidence that VAM association reduces the risk of infection of roots by pathogens in the soil. In return, the fungus obtains sugars as an energy source from the plant.

VAM associations are abundant in angiosperms; most species of angiosperm investigated have VAM root structures. It is probable that many tropical species may not be able to grow in the leached, nutrient-poor soils without such attachments. Mosses and horsetails have also been found with VAM, but the association is rare in gymnosperms.

A number of gymnosperm and angiosperm trees have a different type of fungal symbiosis with short stubby roots surrounded by a thick web of fungal hyphae. The hyphae form a network between the cells of the outer root, but do not enter root cells. These are **ectomycorrhizal** associations. Again, the tree obtains more phosphate, a more efficient water supply and protection from root pathogens. The fungus (usually a basidiomycete, but sometimes an ascomycete) obtains sugars for growth.

2.5 Viruses

Most biologists argue that viruses are not living organisms in the way we understand life. They are not classified within the five kingdoms and are not given binomial names. Instead viruses have abbreviations of their common names.

Viruses are unusual and rather puzzling objects. They cannot produce their own energy or reproduce without first invading a living cell. Their structure is not cellular, not even like a very simple cell: they consist only of one or a few strands of nucleic acid inside a protein coat.

Viruses may be rogue fragments of nucleic acid which escaped from organisms in the past and became independent. Dawkins, in *The Selfish Gene*, emphasises the enormous power of the genes to reproduce, and that the only real function of organisms is to allow their genes to multiply and continue down the generations. Viruses have gone furthest towards expressing this concept of the power of genes to be copied and multiplied: they are genes that have left the confines of an organism and can exist independently in the environment. Although viruses have abandoned the cellular life, they still need the structures in the living cells of a host organism in order to multiply. In doing so they destroy the cell and cause symptoms in the host. Viruses therefore are like **intracellular parasites** of a most insidious kind; many cause severe diseases in plants and animals. Human diseases caused by viral infection include colds, influenza ('flu'), measles and AIDS.

Viruses are remarkably host-specific because the viral coat must interact with the cell surface to enter and because its genes have to trigger copying in the host cell. Often the virus only attacks cells in particular tissues or organs in the host. Most plant viruses contain **single-stranded RNA**, but the cauliflower mosaic virus (CaMV) has **double-stranded DNA**. Some animal viruses have **double-stranded RNA** and an enzyme capable of copying the packaged template genes (see page 35).

A virus outside its host consists of genetic material in a protein capsule (or capsid) and is called a **virion**. Free virions have no locomotory capacity and must be spread by some means. Many plant viruses are spread by insects which suck sap or chew leaves and then transport the virus to a new plant. The flu virion spreads when we cough and splutter. Some are carried by animal vectors. For example, mammals transmit rabies directly by biting the next host. Britain is officially rabies free, although occasionally a rabid animal may cross from the continent. In 1996 a rabid bat was found on the south coast and unfortunately bit two ladies trying to rescue it. Although rabies is fatal, the disease can be prevented in humans by injections, even after a person has been bitten.

When infected by a virus, the mammalian immune system is triggered. This recognises the proteins in the viral coat and produces appropriate **antibodies**. Vaccines which pre-arm the mammal to recognise the virus can give immunity to disease. If the genes producing the coat mutate frequently, the antibodies will fail to recognise a second infection. This is why people catch colds and flu again and again – the virus is a mutation ahead of us!

Tobacco mosaic virus (TMV)

TMV was the first virus to be identified, isolated and studied with the electron microscope. TMV causes mottling of the leaves of the tobacco plant. It is a very simple virus: a single spiral strand of RNA of 6400 nucleotides

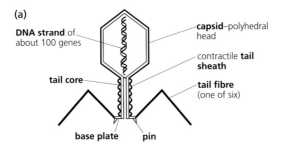

(a)

DNA strand of about 100 genes

capsid–polyhedral head

contractile **tail sheath**

tail core

tail fibre (one of six)

base plate   **pin**

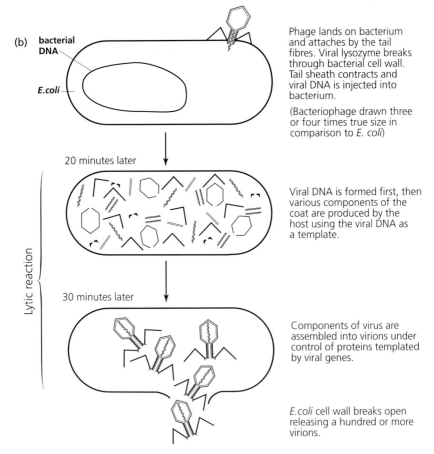

(b) **bacterial DNA**

E.coli

Phage lands on bacterium and attaches by the tail fibres. Viral lysozyme breaks through bacterial cell wall. Tail sheath contracts and viral DNA is injected into bacterium.

(Bacteriophage drawn three or four times true size in comparison to *E. coli*)

20 minutes later

Viral DNA is formed first, then various components of the coat are produced by the host using the viral DNA as a template.

Lytic reaction

30 minutes later

Components of virus are assembled into virions under control of proteins templated by viral genes.

E.coli cell wall breaks open releasing a hundred or more virions.

Figure 2.9 (*a*) The T4 bacteriophage virion is composed of a complex protein coat containing a DNA strand in the head. (*b*) The stages occurring during infection and the lytic reaction by the T4 phage of a host cell of the bacterium *E. coli*. The cell length is about 2–6μm.

inside a rod-shaped protein coat about 300 nm long and 18 nm wide. Early experiments showed that it drained through filters too fine to allow any known bacteria to pass, thus suggesting the presence of a new form of infection. Originally, the word 'virus' meant a slimy poisonous liquid, but as knowledge grew the word came to describe the substance which was extracted from the filtrate, and was finally applied to the tiny particles themselves.

Phages

Oddly enough, some of the most complex viruses are those which invade bacteria. These are called **bacteriophages** or just **phages** (from the Greek *phagein*, 'to eat'). Phages vary in size: some contain only four genes, others as many as 240. In many phages the outer coat is subdivided into head and tail structures; the nucleic acid is usually DNA.

The λ **phage** (λ is the Greek letter lambda) is a parasite of *Escherichia coli*. It has been used, by genetic engineers, to inject particular genetic sequences into strains of *E. coli* (a form of transduction). The bacterium acquires the ability to make certain proteins needed for human health such as insulin and human growth hormone. A similar phage is the T4 phage, also a parasite of *E. coli*; its structure and infection cycle can be seen in figure 2.9.

After the genes of the phage enter the bacterium, two things can happen. In **lytic reactions,** copies of the phage are rapidly produced and the cell bursts (lyses) to release new virions (see figure 2.9b). New phages can be produced in under half an hour. Sometimes though, when the phage enters the cell no immediate replication of the virus occurs. Instead the viral genes join the bacterial genome and are copied only when the bacterium divides. This is called a **lysogenic relationship**. A lytic reaction may follow, triggered by ultraviolet light or a chemical substance.

Human immunodeficiency retrovirus

The human immunodeficiency virus (**HIV**) is responsible for a range of symptoms collectively known as acquired immune deficiency syndrome (**AIDS**). HIV has a polyhedral coat inside an outer, spiky envelope. HIV is unusual in that it contains double-stranded RNA: such viruses are called **retroviruses**. Once inside a human cell, the RNA produces a template DNA strand with the capacity to produce more viral RNA. This is the reverse of the normal copying sequence where DNA is made from RNA templates. HIV virions carry polymerase <u>re</u>verse <u>t</u>ranscriptase to help this process. The letters of the enzyme provide the name <u>retro</u>virus. HIV can remain dormant for years if the DNA template forms a double helix and becomes incorporated in the human genome. HIV is transmitted during very close contact involving transfer of body fluids, namely blood, semen or vaginal secretions, from one person to another.

Viroids

Recently, naked strands of RNA (without protein coats) have been discovered which also invade cells. These have been called **viroids**. Other RNA molecules can only invade in the presence of a virus; these are called **virusoids** and may be fragments of viral RNA which have become separated from the virus but are still functional. Some RNA strands (called **satellites**) seem to modify the effect of the virus and may be a response from the host to viral infection.

Box 2.2   Prions

Bovine spongiform encephalitis, BSE for short and otherwise known as mad cow disease, causes distressing symptoms in cows. A similar disease, called scrapie, has been known in sheep since the eighteenth century. There is strong suspicion that a form of scrapie was passed to cows when the latter were fed with cattlefeed made from the carcasses of sheep. Other mammals in zoos and mink farms, and domestic cats, have recently developed similar diseases. This has led to public concern over eating beef because BSE is similar to a very rare affliction of humans known as Creutzfeldt-Jacob disease (CJD).

Small glycoproteins known as **prions** are responsible for BSE and scrapie. Prions are found naturally in the brain, but their function is unknown. Some strains of mice have now been bred with no prions, yet the mice appear quite healthy. Disease-causing prions are slightly different from normal prions and attack the nervous tissue. How a protein can cause a transmittable disease without the aid of nucleic acid is still unclear but the disease form of the prion appears to convert or modify normal prions into more disease forming prions.

A new variant of CJD, which affects younger people (under 40 years old), has recently appeared. It is similar to the disease in cats and to BSE and may, therefore, be caused by the BSE prion.

THREE

Plants

Plants are crucial for the diversity of terrestrial life. They are photosynthetic autotrophs which form the base of the food chain for all land animals. Trees can grow extremely tall, so forests provide food, cover, nesting sites and protection for many other organisms. It is difficult to imagine land surfaces without plant cover; certainly they would be barren and hostile places with very little biodiversity.

3.1   Colonisation of land

The first land plants

The invasion of dry land by plants occurred relatively late in the earth's history. It probably happened only when enough oxygen had built up in the atmosphere to provide an ozone shield blocking out dangerous ultra-violet light. Water filters out ultra-violet light, so aquatic organisms were already protected.

The first cyanobacteria inhabited shallow shelf areas near the interface between sea and land. Later, algae also occupied near-shore habitats, estuaries and freshwater sites. By the beginning of the Silurian (440 million years ago), the seas contained unicellular plankton, seaweeds, invertebrates and fish, but the land was still relatively bare. Unicellular and filamentous green algae had probably colonised stable and damp surfaces, but dry or hot continental interiors would have been bare rock, sand and drifted dust. Once land plants evolved, a whole series of new habitats became available for animals to colonise.

Adaptations for life on land

Multicellular algae thrive in sea and fresh water, but they are not adapted to living on land. Early land plants evolved many adaptations for survival on hot dry land:
- an outer waterproof layer which prevents dehydration in air – the **cuticle**
- holes in the cuticle which allow oxygen and carbon dioxide through but can be shut when necessary – **stomata**

- anchors in the ground – **rhizomes** (underground stems) or **roots**
- a system for obtaining water and nutrients from the soil – **rhizoids** or **root hairs**
- structures to lift photosynthetic parts above other vegetation – **stems**
- an efficient transport system which carries water, nutrients and photo-synthates – the **vascular system**
- specialised photosynthetic parts which capture light energy efficiently – **leaves**
- protection of reproductive units released into the air – a tough **spore coat**.

Not every early land plant had all these characters: leaves evolved later in some lines, and a vascular system is not found in every fossil. Root evolution is poorly understood as roots do not fossilise well, but rhizomes seem to have evolved first. Once plants colonised the land, competition for light became important: more complex stems evolved which held leaves higher above the ground. The evolution of trees had begun.

3.2   Mosses and liverworts (Phylum Bryophyta)

Some land plants survive without most of the adaptations to life on land listed above. Mosses and liverworts (**bryophytes**) have no vascular system, hardly any cuticle, few stomata and no roots. They thrive on land, but are restricted to moist areas and are the lowliest of land plants – seldom growing more than a few centimetres tall. They are abundant in damp places including bogs, Arctic tundra and on trees in temperate and mountain cloud forests. Many bryophytes are intolerant of pollution, so the presence or absence of particular species indicates local pollution.

Bryophyte structure

Mosses and liverworts possess neither xylem nor phloem. Water, carbon dioxide and minerals are obtained directly from the damp environment across the cell walls. Their chloroplasts contain chlorophyll a and b. Although mosses and liverworts require a moist environment for growth and reproduction, some, which colonise bare rock surfaces, can survive long periods of dormancy in a dry state.

Mosses have narrow stems which grow tightly packed to form cushions with a springy feel. The small, spirally arranged leaves are often only one cell thick (see figure 3.1a). The moss is anchored to the ground by a group of fine multicellular filaments called **rhizoids**.

The large hair cap mosses (including the common British genus *Polytrichum*) have a complex stem with a central column of long cells called **hydroids**, which function as conducting tissue. These mosses form lawns in Japanese moss gardens and some grow over 50 cm tall.

Figure 3.1 Bryophyte structures.

(*a*) The delicate leafy stems of this moss are overtopped by the tough wiry stalks of the sporophyte plants. Each stalk is topped by a green capsule, the   sporangium, containing developing spores.

(*b*) Part of the thallose liverwort *Marchantia*. The little cups on the thallus contain gemmae. These are small vegetatively formed propagules which are dispersed in rain splashes and grow to form new plants.

Other large mosses with spirally arranged side-branches (*Sphagnum*) create raised and blanket bogs in wet climates. *Sphagnum* bogs are rich in other species including ciliates, flagellates, unicellular algae, cyanobacteria, ants, spiders and crustaceans. The mosses absorb basic ions from their watery environment and release hydrogen ions. This acidifies the water (sometimes to pH 3) so that decomposer microorganisms cannot survive and organic peats build up. Peat accumulation at higher latitudes is considerable; indeed, more carbon is locked up in peat than in tropical rainforest. Decomposition of this peat, caused by human activity such as the drainage of bogs, or drying-out triggered by climate change, releases carbon as atmospheric carbon dioxide. If the greenhouse effect increases peat decay, atmospheric carbon dioxide levels will be further increased, thus resulting in **positive feedback.**

Sphagnum peat has many uses. In Britain it is still used for fuel and by gardeners. The dried moss itself is very absorbent and has its uses: for nappies, bedding and cleaning up oil spills. It contains a salve called sphagnol, which is useful for skin diseases and insect bites. *Sphagnum* can also be applied as a dressing over wounds. Peat extraction is a problem as bogs are easily destroyed; in Britain, there are some internationally important ones which are threatened by wholesale damage.

Liverworts are more varied than mosses. Some look like moss, but have leaves in two or three rows and their rhizoids are unicellular. Others have a flat curling form (called a **thallus**) anchored by the unicellular rhizoids (see figure 3.1b). Thallose liverworts often live together with fungi or cyanobacteria.

Life cycle

The bryophyte life cycle shows **alternation of generations** (see box 3.2). The main moss and liverwort plants are **gametophytes** with haploid cells containing only half the full complement of chromosomes. They have organs which produce gametes. The diploid plant, called the **sporophyte** (because it produces spores) is very small and remains attached to the gametophyte.

The gametophyte grows from a spore. First it forms a filamentous green mass called a **protonema,** which matures into the leafy stage. Sexual organs develop: some produce eggs and are called **archegonia**, others produce sperm and are called **antheridia**. Often, archegonia and antheridia develop on different plants. The sperm have flagella and require a film of water in which to swim; they are attracted to exudates produced by a ripe egg. After fertilisation, the diploid zygote remains anchored in the gametophyte. The resulting sporophyte relies on the gametophyte for nutrients and water.

In mosses, the sporophyte grows a tough stalk, which lifts a developing **sporangium** above the gametophyte (see figure 3.1a). Within the sporangium, cells divide by meiosis to produce haploid spores. The sporangium wall hardens into a protective capsule – the only part of a moss with stomata. When the spores are ripe, the capsule splits and a lid flips off, allowing the spores to be gradually shaken out in the wind. The variety of capsules and lid shapes, some with ornate teeth, aids moss identification.

In liverwort sporophytes, the sporangium develops before the stalk elongates. The capsules never have stomata, and break open along four slits. The spores are shed in one flurry, after which the fragile stalk soon disintegrates.

One widespread liverwort group (including *Marchantia*, see figure 3.1b) has antheridia and archegonia which develop at the top of stalks. This aids sperm dispersal, and the sporophyte, because it is already above the thallus,

Box 3.1   Vascular systems

Most land plants have a vascular system, which transports substances around the plant. The system consists of **xylem,** which carries water and nutrients from roots to leaves, and **phloem,** which distributes the products of leaf photosynthesis.

Xylem contains long, empty cells with thickened, lignified walls. Water flow in the xylem is mechanical: water loss (**transpiration**) from the leaves pulls more water up the stem. Most plants have xylem cells with pointed ends called **tracheids.** Many angiosperms also have specialised open-ended cells, which form long tubes called **vessels.** Vessels are characteristic of angiosperms, but a few other plants also have vessel-like tubes, which evolved independently. Vessels conduct water faster than tracheids because of their open tubular structure.

Phloem is a living tissue: when phloem cells die they cease to function. In most plants, elongated **sieve cells,** which lack a nucleus, are interspersed with living **albuminous cells.** In angiosperms the sieve cells form tubes (**sieve tubes**) and the living cells associated with them are called **companion cells.** Solute transport in the phloem is an active process powered by the albuminous and companion cells.

The first (**primary**) xylem and phloem form behind the growing tip of any stem. Many plants have only primary systems, but in some, a band of special cells (the **cambium**) goes on producing **secondary** xylem and phloem throughout the plant's life (see figure 3.2). The build-up of secondary xylem forms **wood,** which strengthens large plants and provides a water store. Phloem only has a short life; secondary phloem is usually crushed each year and replaced from the cambium. Plants with a vascular system are called **tracheophytes** and include ferns, horsetails, clubmosses, gymnosperms and angiosperms.

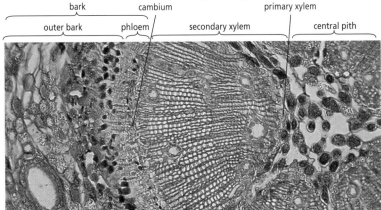

Figure 3.2 Transverse section through a young conifer, Scots pine (*Pinus sylvestris*) stem showing secondary xylem and two annual rings of growth.

does not need its own stalk. In *Marchantia* the sporophytes occur in a nine-armed star. This is not a typical liverwort as it also has a cuticle and pores on the upper surface of the thallus, but it is so striking that it is the one most people have noticed.

3.3   Ferns (Phylum Filicinophyta)

Plants with a vascular system and which also produce spores include ferns, horsetails and clubmosses and are called **pteridophytes**. Ferns (Filicinophyta) are the largest group within the pteridophytes.

Structure

Fern leaves are usually large and divided into leaflets (they are **pinnate**); they grow in a crown at the top of the stem. Some leaves develop spore-bearing organs on their undersides. In the tropics and subtropics, tree ferns 10 to 20 m tall form an important part of the forest.

Fern stems have complex primary vascular systems, with strands running out into the leaves. Leaf development is very distinctive with spirally curled young leaves (crosiers or fiddleheads), which unfold as they grow. Large stems are often supported by protective leaf bases and adventitious roots, which grow out of the stem in a tangled mass. Some ferns are **epiphytes** attached to tree trunks or branches. Their leaves often form a cup in which drifting debris rots down and provides nutrients. Tropical ferns are sometimes sold in garden centres as house plants: the bird's nest fern and stag's horn fern are epiphytic examples.

A common British fern is bracken (*Pteridium aquilinum*). Even though it is not an angiosperm, bracken has vessels (see box 3.1). It occurs throughout the world and is often a persistent weed. It spreads by horizontal underground rhizomes. These are difficult to kill because they are too deep to dig out or burn and are resistant to many herbicides. Bracken, if eaten, is toxic to most animals and carcinogenic to humans.

Life cycle

Unlike in bryophytes, it is the fern sporophytes which are large and leafy. Spores develop in clusters of sporangia on the undersides of leaves (see figure 3.3). The spores are the product of meiosis and, after they are released, germinate and grow into haploid gametophytes. The gametophyte, which looks like a small green corn flake, develops archegonia and antheridia.

Flagellated sperm are released from the antheridia after rain and swim in water films. Usually the antheridia on a gametophyte mature before the archegonia, so self-fertilisation is prevented and outbreeding occurs. The resulting zygote grows into a new sporophyte. Some gametophytes

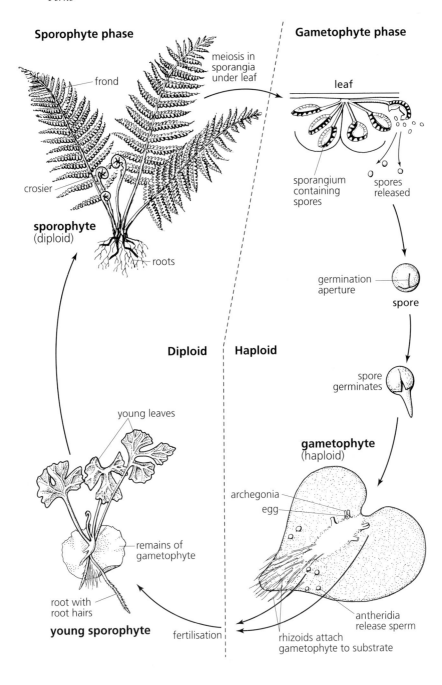

Figure 3.3 The reproductive cycle of a typical fern.

Box 3.2   Alternation of generations

Alternation of generations is a descriptive term for the way many plants have two different alternating life forms: one haploid (the gametophyte) and the other diploid (the sporophyte).

In chapter 2, we skipped the complexities of algal reproduction. In multicellular algae, the sporophyte and gametophyte are usually similar in size and longevity: a reproductive cycle with equal emphasis on the diploid and haploid generations.

In bryophytes, this balance is shifted in favour of the gametophyte. The long-lived moss and liverwort plants are haploid gametophytes and the diploid sporophyte is reduced to a small holdfast, stalk and sporangium attached to, and relying on, the gametophyte. This suggests that bryophytes are a side branch in land plant evolution. In most lines of plant evolution the trend is for the gametophyte to become more and more reduced. Hence in ferns and other pteridophytes (horsetails and clubmosses), the sporophyte is large and the gametophyte reduced and, usually, short-lived.

In seed plants (gymnosperms and angiosperms), the gametophytes are even more reduced and never live independently in the environment because they are contained within the body of the sporophyte. The fossil record shows that in early plants an evolutionary trend resulted in sporophytes which produced spores of two different sizes: small microspores and large megaspores. Both types were released into the environment. The microspores germinated to produce gametophytes which developed antheridia, while the larger megaspores germinated to produce gametophytes which developed archegonia; the gametophytes were thus either male or female. Very few species with microspores and megaspores now survive.

In some fossil species, the megaspore was very large and remained attached to the sporophyte, protected from animals by tough scales. The megaspore coat burst open to reveal the archegonia. Microspores landed on the megaspore surface, where the male gametophyte germinated and released sperm to fertilise the eggs.

Through evolution, the megaspore, enclosed in a spore wall, has become completely encased in sporophytic tissue, resulting in a structure called an **ovule**. The microspores – called **pollen** in gymnosperms and angiosperms – settle on specialised landing structures of sporophyte material and germinate to grow towards the female gametophyte. They release sperm, motile in some taxa though not in others, which fertilise the ripe eggs. Once fertilised, a zygote develops into an embryo. The embryo remains encased in remnants of the gametophyte and outer protective cases of the parent sporophyte: this whole structure is called a **seed**.

Gymnosperm means 'naked seed' and indicates an incomplete enclosure of the gametophyte by sporophytic tissue. An angiosperm ('vessel seed') has a completely enclosed gametophyte; the male gametophyte has to grow through sporophytic tissue to reach it.

---

proliferate by vegetative budding before they produce sex organs. A few gametophyte strains are known which never produce sex organs and just divide vegetatively. These unusual forms are stuck in the one life form and seem to have lost alternation of generations.

3.4 Gymnosperms

The gymnosperms are an ancient group of seed plants. Their long evolutionary history is reflected by the very different taxa in the group. Many extinct gymnosperm fossils (e.g. the cycadeoids, see box 7.1 on page 112) are found from geol- ogical periods when gymnosperms dominated the world's vegetation, before angiosperms evolved. The main two groups living today are conifers (Phylum Coniferophyta) and cycads (Phylum Cycadophyta). Cycads have long, tough, pinnate leaves, which grow in a crown at the top of a thick trunk. In some, the trunk grows underground and just the leaves emerge from the soil. The cycads are a tropical and subtropical group.

Another noteworthy gymnosperm is the maidenhair tree (*Ginkgo biloba*), a broad-leaved deciduous tree with slimy green fruits the size of acorns, each of which contains a single seed. *Ginkgo*-like leaves are abundant in the fossil record of 200 million years ago. A few living trees were discovered in a Buddhist monastery garden in China in the 1750s, so *Ginkgo* is only known in cultivation. It is a 'living fossil', the sole survivor of a once abundant phylum.

Another small phylum, the Gnetales, is of interest for its amazing diversity and unusual morphological range. It includes bushy plants of cold semi-deserts, a series of lianas which grow in tropical forests and have broad leaves indistinguishable from angiosperms, and a weird plant with an underground stem like a giant turnip. This last plant lives in the Namib desert and has only two leaves like long scarves. Gnetales all have vessels in their vascular systems and you would have great difficulty recognising them as gymnosperms!

Conifer structure

Conifers are all trees with small scale- or needle-shaped leaves (see figure 3.4). Most conifers are evergreens with tough, thick leaves, which have to survive and function for several years on the tree. When the leaves fall, their tough structure prevents immediate decay and coniferous forests are noted for the deep carpet of leaves on the floor. Their secondary xylem is constructed of tracheids (see figure 3.2).

Conifers are rare in the tropics; most grow at higher latitudes in temperate and boreal habitats. They are a major tree type at high altitude, in the Alps and Andes, for example. Many conifers are cultivated for their wood, which is used for building, carpentry and paper. The demand for

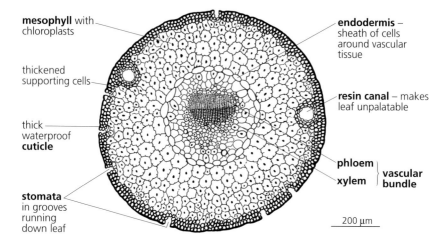

mesophyll with chloroplasts

endodermis – sheath of cells around vascular tissue

thickened supporting cells

resin canal – makes leaf unpalatable

thick waterproof **cuticle**

phloem | **vascular**
xylem | **bundle**

stomata in grooves running down leaf

200 µm

Figure 3.4 A section through a needle leaf of the conifer *Pinus monophylla*.

wood results in conifer plantations, often damaging other important habitats like upland moor and bogs.

Conifer reproduction

The reproductive structures in conifers, as their name suggests, are arranged in **cones**. Cones are spirally arranged modified leaves (cone scales), each carrying either pollen sacs or ovules. Male cones are usually smaller than female cones. The familiar British yew (*Taxus baccata*) looks like a conifer, except that is has a single seed surrounded by a fleshy red coat instead of a female cone. Some taxonomists separate it from conifers because of this.

The reproductive cycle of a typical conifer is illustrated in figure 3.5. Pollen is released from the male cones in vast amounts and blows around on the wind. Pollen only lands on a female cone by chance. An isolated tree surrounded by trees of other species seldom has its ovules fertilised. This limits conifers to living in groups of the same species.

In conifers the pollen tube grows towards the megaspore, and sperm are released directly at the neck of the archegonium. (In cycads and *Ginkgo* the gametophyte releases motile sperm, which swim towards the archegonia in the encased megaspore.) After pollination, the cone closes and seeds develop, protected by the tough cone scales. Ripe cones re-open and the seeds, with papery wings, are dispersed on the wind. Some cones remain closed, the seeds only being released when animals break open the cones for food. For example, the nutcracker bird opens pine cones in the Alps. In some conifers, the cone opens only after fire, and the seeds are released into a burned habitat with no competition from tall plants.

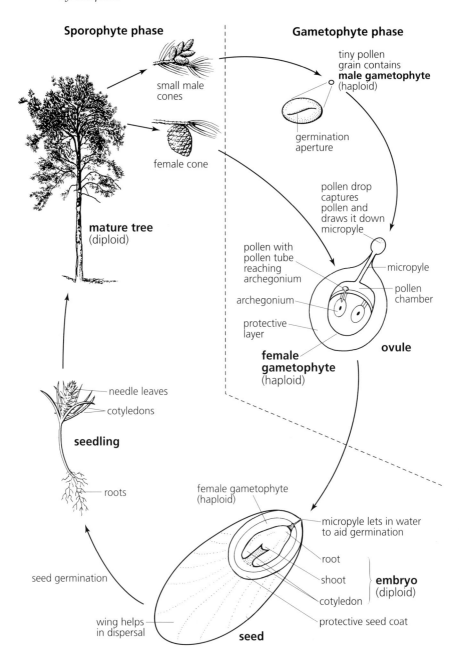

Figure 3.5 Reproductive cycle of a typical conifer.

3.5   Angiosperms

Flowering plants (angiosperms) have triumphed in terrestrial habitats: they occur everywhere except hot and polar deserts. Many live in freshwater lakes and rivers and a few even live in the sea. Their diversity seems mind-blowing, as do the complex relationships they have developed with other organisms, especially insects, mammals and fungi. Angiosperms were the last major plant group to evolve. They appeared in the fossil record about 120 million years ago, and rapidly increased in diversity and abundance. The first angiosperms probably evolved from woody gymnosperms, but took on many new life forms not seen in gymnosperms. Non-woody angiosperms with no secondary xylem are abundant and dominate grass-lands and alpine meadows. Many are short-lived annuals and biennials or survive drought as underground roots and bulbs.

There are two groups of angiosperms: the **Monocotyledoneae** and the **Dicotyledoneae** (monocots and dicots for short). These names are based on one difference between them: the number of leaves (called **cotyledons**) in the embryo. However, there are several differences between monocots and dicots:

- Monocots have one cotyledon in the embryo; dicots have two.
- Monocot flowers have petals and sepals in threes or multiples of three; dicots in fours or fives.
- Monocot leaves are often narrow and strap-shaped with parallel veins; dicot leaves are often broader and much more variable with reticulate (net-like) patterns of veins.
- Monocots are herbaceous and lack secondary xylem, and tall forms (e.g. palms) are usually unbranched, without wood in their trunks; dicots include herbaceous and woody types.
- Monocot vascular systems have many scattered bundles; dicot vascular systems form a ring in cross-section.
- Monocot pollen grains often have one aperture; dicot pollen usually has three or more apertures, although some dicots, thought to be ancient groups, have one aperture.

Of course, these differences are generalisations – the vast diversity of angiosperms provides exceptions to every 'rule'. To record all the variations in morphology would require several more books. Here is a brief look at some 'typical' angiosperms and how they function.

Leaf structure and photosynthesis

Most angiosperm leaves are much larger than conifer leaves. They have a flat **lamina**, strengthened by **veins,** which carry the leaf's vascular system. The leaves are held up by stems and branches so that they can intercept sunlight. Angiosperm leaf structure is adapted to maximise photosynthetic efficiency. A section through a leaf (see figure 3.6) shows the structure and position of the closely aligned **palisade cells** containing chloroplasts. Each chloroplast contains numerous membranes, stacked into **grana,** where photosynthesis occurs.

Beneath the palisade are loosely arranged **spongy mesophyll cells,** with air spaces connecting to stomata in the lower epidermis. When the stomata are open, during the day, carbon dioxide diffuses into the air spaces and released oxygen diffuses out. At the same time, water vapour is lost from the leaf, creating a transpiration flow through the xylem from the roots.

Desert plants have limited water reserves, so it is damaging for them to open their stomata during the day. Many open the stomata only at night, when it is cool, so that transpiration is minimised. Carbon dioxide is taken into the leaf at night and stored in large cell vacuoles as malic acid. Stomata shut in the daytime, but the plant has a carbon dioxide reserve in the malic acid, which is released as required. These plants are called CAM plants (CAM is short for Crassulacean acid metabolism, named after the plants in which it was discovered). The large vacuoles make CAM plants succulent. The vacuoles not only act as malic acid stores, but also provide a heat sink preventing the plant overheating during the day when it cannot be cooled by transpiration.

Flower structure and function

Showy, bright-coloured flowers are the most distinctive feature of many angiosperms. A simple flower is illustrated in figure 3.7a. This flower is **radially symmetrical** – typical of buttercups and geraniums, but there are many modifications of this pattern. In tulips the sepals are the same size and colour as the petals (see figure 3.7b). In orchids the petals and sepals are different sizes and shapes, producing a flower with **bilateral symmetry** (see figure 3.7c). In some flowers the petals are fused into a long tube. All these

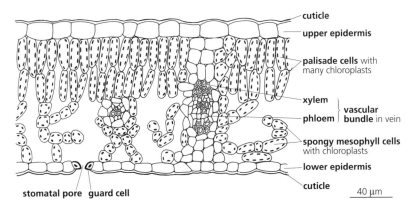

Figure 3.6   A section through a typical dicot leaf. (Redrawn from Gifford E. M. & Foster A. M. 1989 *Morphology and evolution of vascular plants,* W. H. Freeman.)

adaptations are designed to attract animals such as insects, birds and bats to the flowers, where they collect sugary nectar for themselves and, in doing so, accidentally pick up pollen. When they visit another flower the pollen is rubbed off onto the stigma and the flower is pollinated. The general term for animal-pollinated flowers is **zoophilous** flowers; those pollinated by insects alone are **entomophilous** (see box 5.2 on page 80).

Some angiosperm flowers release pollen in large quantities into the air. As they do not attract animals, the petals and sepals are small and not brightly coloured. These are **anemophilous** (wind-pollinated) flowers and include grasses (see figure 3.7d) and willows.

The form of the flower (number of petals, sepals and stamens, position and structure of the ovary, arrangement and number of ovules and arrangement of flowers in a flowerhead) is used to classify plants into families. Because flower structure is so variable, a confusing number of descriptive terms are

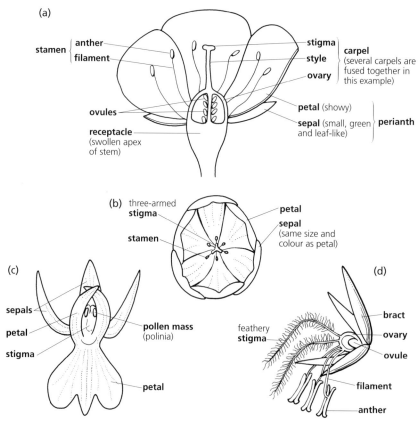

Figure 3.7 (*a*) A typical insect-pollinated dicotyledonous flower. (*b*) A tulip flower from above showing the three petals and identical sepals. (*c*) An orchid flower. Despite the complex shape, the typical three petal and sepal arrangement of the monocots can still be seen. (*d*) A wind-pollinated grass flower showing the long feathery stigma for catching airborne pollen.

used. In figure 3.7 just the basics for naming parts of the flower are given.

As in gymnosperms, the pollen grain contains the male gametophyte. When pollen reaches the receptive female parts of a flower, it lands on a special stigmatic surface. It is still a long way from the ovules, which are in a chamber encased in sporophytic tissue. The pollen grain germinates by bursting open and a pollen tube (the gametophyte) grows out and down through the sporophytic tissue to reach the ovule and fertilise the egg.

During germination and growth through the flower tissue, the pollen tube is very vulnerable. It can be prevented from reaching the ovules by the sporophyte if the pollen is incompatible. The most obvious incompatibility is if the pollen is of a different species and has landed by accident on the stigma. Usually such pollen never germinates. Pollen may also be incompatible if it comes from the same individual. In some species, the sporophyte reacts to proteins on the pollen surface and prevents the pollen tube from growing down through the stigma. The proteins are added in the anther during pollen maturation, so they come from the pollen parent. If they are recognised by the sporophyte as being too similar, then pollen tube growth is blocked (**sporophytic incompatibility**).

In other species, substances from the pollen tube itself are detected by the sporophyte. Again, if these are recognised as being similar by the sporophyte, tube growth is blocked (**gametophytic incompatibility**). In both sporophytic and gametophytic incompatibility, the reactions prevent self-fertilisation of ovules by pollen from the same plant, or even fertilisation by close relatives: they promote **outbreeding**. Such selection mechanisms by the sporophyte may have been important in promoting a rapid diversification of angiosperms. Other systems which favour outbreeding occur. Some species produce male or female flowers on separate plants: they are **dioecious** (kiwifruit and holly are examples).

Understanding plant reproduction is important in the production of food crops. For example, some apples will not set fruit unless a different strain provides the pollen. Dioecious species like the kiwifruit need a few males in a field of females to provide the pollen. Understanding how pollen reaches a plant is also important: sweetcorn and wheat, for example, are wind-pollinated, so they are best grown in monoculture where the pollen can easily float to another plant. Insect-pollinated crops may need native vegetation close by to attract native pollinators such as bumble-bees. Many crops have honey-bee hives put in them to aid pollination. But some flowers cannot be pollinated by honey-bees, which do not shake the flower enough to release the pollen; tomato and kiwifruit are examples. Preserving wild bumble-bee populations is therefore important for pollination of some crops.

Invertebrates other than arthropods

This and the next two chapters are about animals. In terms of numbers of species, the animal kingdom is the largest of the five kingdoms. It contains about 30 phyla, but we shall consider only 8: cnidarians, platyhelminths, nematodes, annelids, molluscs and echinoderms in this chapter; arthropods in chapter 5; and chordates, the phylum to which we belong, in chapter 6.

The chordates include the **vertebrates** – animals with backbones. All the other phyla listed above are **invertebrates** – animals without backbones. Invertebrates comprise 97% of all living species of animals, and they show considerable diversity in structure and function. We shall start with the simpler ones and then go on to more complex forms.

4.1 Cnidarians (Phylum Cnidaria)

This phylum includes *Hydra*, jellyfishes, sea anemones and corals. Although *Hydra* occurs in fresh water, most cnidarians are marine. None lives on land.

Cnidarians have a simple body like a sack (see figure 4.1a). The mouth opens into a large digestive cavity called the **enteron**. There is **no anus**; indigestible waste is ejected through the mouth.

The enteron is surrounded by the **body wall**, which is composed of only two layers of cells: the **ectoderm** faces outwards, forming the 'skin', and the **endoderm** faces inwards, lining the digestive cavity. Between the ectoderm and endoderm is a non-cellular layer of gelatinous material called the **mesogloea**. Cnidarians and related animals whose body wall is made up of two layers of cells are described as **diploblastic**.

Seven different types of cell, each with one or more specific functions, are found in the body wall of a cnidarian such as *Hydra* (see figure 4.1b). So there is some degree of cell differentiation with a division of labour between the cells. Certain of the cells in the body wall carry out more than one function. Take the **musculoepithelial cells**, for example: the flat outer part of the cell forms, with its neighbours, a protective **epithelium** while the elongated inner part is a contractile **muscle fibre**. Such unspecialised cells are characteristic of cnidarians; most other animals have separate epithelial and muscle cells.

One type of cell, though, is remarkably specialised – the 'sting cell'

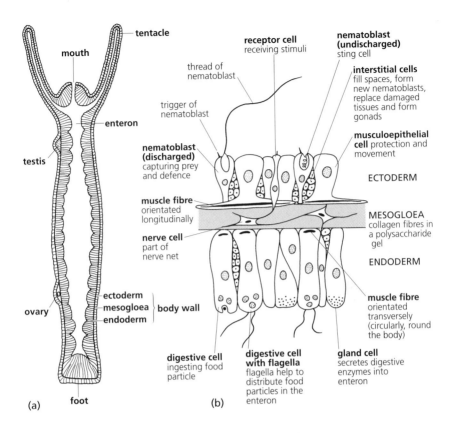

Figure 4.1 Longitudinal section through the body wall of *Hydra* showing (*a*) the diploblastic structure and (*b*) the different types of cells in the two cell layers. Digestion starts in the enteron. Small particles of the partially digested food are then taken up into the digestive cells where digestion is completed.

(**nematoblast**). These cells, unique to cnidarians, produce a poison which in some species is highly toxic. The poison of the Australian 'stinger', a small jellyfish, is as toxic as that of the Indian cobra and can kill a person in less than two minutes.

Nematoblasts occur in clumps on the **tentacles** which, thus armed, are used for defence and catching prey (mainly water fleas in the case of *Hydra*). Each nematoblast has its own built-in triggering mechanism and is independent of the nervous system. When the trigger is touched, a long thread like a harpoon shoots out and pierces the prey. Other types of nematoblast are non-toxic and have sticky or coiled threads for clinging to the prey. Once it has discharged its thread, a nematoblast cannot be used again and has to be replaced. The tentacles pull the food to the mouth, through which it is ingested into the enteron.

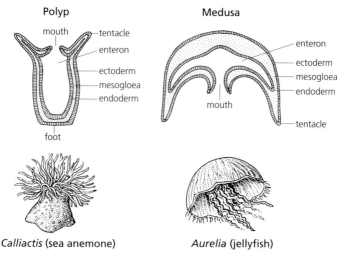

Figure 4.2   Polyp and medusa compared, with an example of each.

Cnidarians have two body plans: the **polyp** and the **medusa** (see figure 4.2). The polyp consists of an elongated column with the mouth and tentacles at the top and a **foot** at the bottom. The foot is usually attached to rocks or weeds, so most polyps are **sessile**. Hydras and sea anemones are polyps. The medusa is a broad, upside-down version of the polyp and is usually free-swimming. It floats in the water with its mouth and tentacles hanging downwards, its shape and buoyancy maintained by the gelatinous mesogloea, which is much more extensive than in polyps. Jellyfishes are medusae.

The polyp and medusa are both **radially symmetrical**, a type of symmetry typical of cnidarians. The various structures (such as the tentacles) are disposed round a central point like the spokes of a wheel. Radially symmetrical animals tend to be sessile or semi-sessile, and when they move there is no obvious leading end. We cannot identify anterior and posterior ends, left and right sides, or dorsal and ventral surfaces. But we can make a distinction between the end that has the mouth (the **oral end**) and the other end.

Cnidarians are essentially simple in structure. Apart from the tentacles and gonads (ovaries and testes), there are **no organs**. The body wall can best be described as a multipurpose tissue, for which reason these animals are said to show the **tissue level of organisation**.

Their physiological processes tend to be simple too. For example, gaseous exchange takes place by diffusion all over the body surface. Being diploblastic, all the cells are close to the surface so the distance through which gases have to diffuse (the **diffusion distance**) is short. There are no special gaseous exchange surfaces, nor is there a circulatory system. Jellyfishes have a system of ciliated water-filled canals for distributing food particles, but otherwise transport systems are unknown in cnidarians.

Locomotion and response

Muscle fibres occur in both the ectoderm and the endoderm. Since they can contract, they enable cnidarians to move and change shape. They are attached to the mesogloea which, being elastic, gives the body flexibility.

The muscle fibres are coordinated by a **nerve net** located between the ectoderm and endoderm. The interconnected nerve cells link the muscle fibres with receptor cells sensitive to touch and chemicals. The nerve net transmits impulses very slowly by human standards, but we have here the basic components of a reflex system with receptors, nerves and effectors.

Reproduction and life cycle

Cnidarians reproduce asexually by **budding**: new individuals grow out from the side of the parent, break away and become self-supporting. Sexual reproduction also occurs: typically, eggs and sperm are released from ovaries and testes into the surrounding water where fertilisation takes place (**external fertilisation**). The zygote then gives rise, directly or via a larval stage, to a new individual.

A typical cnidarian has a life cycle in which a sessile, asexually reproducing polyp alternates with a free-swimming, sexually reproducing medusa. Different cnidarians vary in the emphasis placed on the polyp and medusa. In *Hydra* and sea anemones the polyp is the dominant phase in the life cycle (in fact *Hydra* has no medusa at all), whereas in jellyfishes the medusa is the dominant phase.

Colonial cnidarians

Some sessile cnidarians, such as *Obelia*, consist of clusters of hydra-like polyps linked together in a **colony**. Related species, such as the Portuguese man-of-war (*Physalia*), are complex floating colonies with polyp and medusoid individuals carrying out different functions, so there is a division of labour within the colony (see figure 4.3).

Figure 4.3 The Portuguese man-of-war, *Physalia*, a complex floating colony of polyp and medusoid individuals. Attached to the underside of the 'float' are numerous polyps and medusae. The medusae produce eggs or sperm and are for reproduction; the polyps are for catching prey and feeding. The prey-catching polyps (which are also used for defence) lack mouths but have contractile tentacles which may hang down more than 7 metres and are armed with batteries of nematoblasts.

Other colonial cnidarians, related to sea anemones, are reef-building **corals**. Numerous interconnected polyps secrete a mass of calcareous cups into which the tentacles can be withdrawn for protection. The shape and form of the coral varies from one species to another.

The aggregation into a colony of different types of individual, each specialised to carry out a specific function, is called **polymorphism**. Complex cnidarians, like the Portuguese man-of-war, have carried polymorphism to the point where the colony may be regarded as a single 'super-organism', the whole assemblage behaving as an integrated entity with the different individuals functioning as 'organs'.

4.2   Platyhelminths (Phylum Platyhelminthes)

Platyhelminths (flatworms) occur in fresh water, the sea and on land. Some of them (flukes and tapeworms) are parasites. Free-living flatworms include planarians, which live under stones in fresh water, and they illustrate the basic characteristics of this phylum (see figure 4.4).

The body is flat and there is a mouth but **no anus**. In contrast to cnidarians, flatworms are **bilaterally symmetrical**. In this sort of symmetry most of the structures are arranged symmetrically on either side of a line drawn down the centre (the midline) (see figure 4.4a). A bilaterally symmetrical animal has an **anterior** and a **posterior** end, **dorsal** and **ventral** surfaces, and left and right sides. Bilateral symmetry is associated with motility, and the anterior end is normally the leading end.

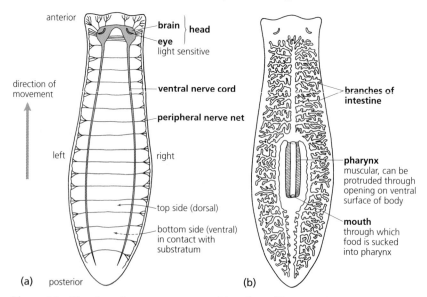

Figure 4.4   The planarian nervous system (*a*) and gut (*b*). Notice how this animal is bilaterally symmetrical.

An important feature of flatworms is the presence of a third layer of cells, the **mesoderm**, between the ectoderm and endoderm. Animals with three layers of cells are described as **triploblastic**. Triploblastic organisation is a fundamental characteristic of the vast majority of animals including humans.

The mesoderm is associated with another important feature, the development of **organs**. Most internal organs are formed from the mesoderm, so they lie between the ectoderm and endoderm. However, a flatworm's organs are relatively simple and, with the exception of the reproductive system, there are no organ *systems* of the kind found in more complex animals. In fact, flatworms do things rather simply. For example, they rely on diffusion for gaseous exchange and internal transport; this is made possible by their flatness, which gives them a high surface–volume ratio and reduces the diffusion distance.

As you can see in figure 4.4a, the nervous system is more elaborate than that of cnidarians. There is a nerve net and, in addition, a pair of **ventral nerve cords**. The nerve tissue in the head is concentrated to form a rudimentary **brain**. It represents the first glimmerings of a **central nervous system**.

Free-living flatworms move away from light. They have simple light-sensitive **eyes** on the dorsal side of the head, and scattered sensory cells sensitive to touch and chemicals are more concentrated in the head than elsewhere. This is what you would expect at the leading end of the body. The tendency in evolution for sense organs, nerve tissue and feeding devices to become concentrated at the anterior end, resulting in the formation of a **head**, is called **cephalisation**. Flatworms show an early stage in this process.

Curiously, though, a planarian's mouth is not at the head end. It is on the ventral side, about two-thirds of the way down the body (see figure 4.4b). Food is sucked into it by means of a muscular **pharynx**. Flukes, however, have their mouth and pharynx at the anterior end. In complete contrast, tapeworms have no gut at all. They inhabit the small intestine of humans and other mammals where they absorb the host's digested food by diffusion across the body surface.

In planarians and flukes, the pharynx opens into an **intestine** from which numerous blindly ending branches ramify to all parts of the body. The soluble products of digestion diffuse directly from these branches to all the cells. A circulatory system is therefore not required.

The urinary system is also spread through the body. It consists of numerous **flame cells** which, aided by the beating of a bunch of flagella, collect unwanted water, salts and waste substances and expel them, via a system of tubes, to the exterior.

If all this seems rather simple, flatworms have one system that is surprisingly complex: the reproductive system. The animals are **hermaphrodite** with male and female organs in the same individual. The arrangement of the organs, and the mating process, ensure that self-fertilisation does not normally take place.

Box 4.1 To be a parasite

Living inside another organism may seem an easy way of life – a comfortable environment and a ready source of food: what more could one want? But there are problems. How, for example, do you prevent yourself being attacked by your host's immune response (or by its digestive enzymes if you live in the intestine), and how do your offspring get to new hosts? Being a parasite is, in fact, full of problems.

These problems are overcome by a number of special features which parasites possess. These **parasitic adaptations** are particularly well illustrated by the blood fluke *Schistosoma* whose main (primary) host is the human (see figure 4.5).

The slender form of the adult worm suits it for living inside its host's veins. It is protected from the host's immune response by coating its cuticle with the host's own molecules, so the host mistakenly recognises it as 'self' and does not launch antibodies against it. This trick is also used by other parasitic flatworms and roundworms.

To raise the chance of completing its life cycle, vast numbers of offspring are produced. Within the primary host, the male and female are clamped in permanent copulation, and a single female may lay as many as 3500 eggs in a day. Each egg develops into a free-swimming ciliated larva called a **miracidium**. Further reproduction (asexual) takes place inside an intermediate (secondary) host, a snail, resulting in the formation of numerous further larvae (cercariae). In the course of a few months, over 200 000 cercariae may be derived from a single miracidium.

The miracidium larva gains entry into the soft foot of a snail by secreting a protein-digesting enzyme from glands at its pointed anterior end. The cercaria is even better equipped for gaining entry into a human. A muscular sucker at the anterior end attaches itself to the skin. Then enzymes (including hyaluronidase, which digests connective tissue) pour out from glands that open close to the sucker.

Alternate shortening and lengthening of the body, together with vigorous movements of the tail, enable the cercaria to work its way through the epidermis into the softer tissue beneath. Entry takes from five to ten minutes altogether.

The parasite then migrates through the lymphatic vessels and blood system to the liver, where it matures. After this, it moves into the branches of the hepatic portal vein. An adult worm may live for as long as 30 years, and a single host may have several thousand of them.

Schistosomiasis (bilharzia), the disease caused by the blood fluke, is debilitating but not immediately fatal. In fact, a person may live with it for years. This, too, is a parasitic adaptation, for a parasite that quickly kills its host destroys itself as well.

Parasitism involves the exploitation of one species by another and is extremely widespread. Bacteria, protoctists and fungi parasitise animals

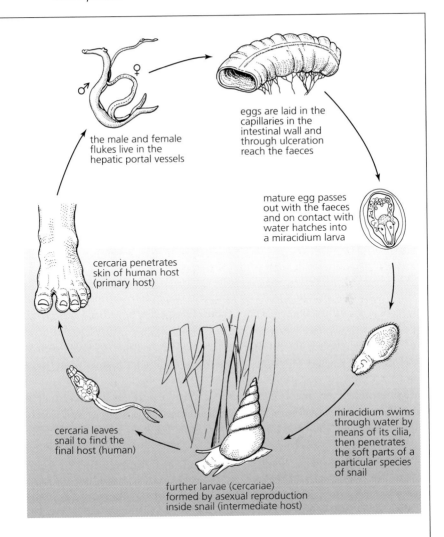

the male and female
flukes live in the
hepatic portal vessels

eggs are laid in the
capillaries in the
intestinal wall and
through ulceration
reach the faeces

mature egg passes
out with the faeces
and on contact with
water hatches into
a miracidium larva

cercaria penetrates
skin of human host
(primary host)

miracidium swims
through water by
means of its cilia,
then penetrates
the soft parts of a
particular species
of snail

cercaria leaves
snail to find the
final host (human)

further larvae (cercariae)
formed by asexual reproduction
inside snail (intermediate host)

Figure 4.5 Life cycle of the blood fluke *Schistosoma japonicum*, which occurs in the Far East. Similar species occur in Africa and South America. (Adapted from Noble & Noble 1964 *Parasitology*, 2nd edition, Lea and Febiger.)

and plants; animals parasitise plants and other animals; plants parasitise other plants; and parasites parasitise other parasites. A single host species may be parasitised by more than one type of parasite. For example, *Calanus*, a small crustacean, may be parasitised by a fluke, a tapeworm, a roundworm, an isopod, a fungus, a sporozoan, a ciliate and several species of flagellates. Parasitism is therefore an important aspect of biodiversity, and the perspective is broadened even further if we take into account other associations such as mutualism (see page 14).

Platyhelminths include some notorious parasites. Certain flukes are **ectoparasitic**, living attached to, for example, the skin or gills of fishes and amphibians. They have **suckers** and **hooks** for clinging to the host. Other flukes, and all tapeworms, are **endoparasitic**, living inside the tissues or gut of various hosts including humans. Those that feed on the tissues can be very harmful. For example, sheep infected with the liver fluke become thin and weak and will die if remedial action is not taken.

Some endoparasitic flatworms have complex life cycles with more than one host species, and this is exemplified by the blood fluke. The life cycle of this animal, and the reason for its complexity, are discussed in box 4.1.

4.3   Nematodes (Phylum Nematoda)

Nematodes (roundworms) occur almost everywhere. Many are **saprobionts**, feeding on dead organic matter. A spadeful of garden soil may contain as many as a million of them. Twenty thousand species have been described, but the total number could be as high as half a million.

A nematode has an elongated cylindrical body, round in cross-section (see figure 4.6) – hence the name 'roundworm'. There is an **anus** as well as a mouth, and the gut is a straight tube without branches of the kind found in flatworms. There are no special gaseous exchange surfaces and no circulatory system. Their small size and long thin form gives them a sufficiently large surface–volume ratio for gaseous exchange to take place by diffusion across the surface of the body.

The body surface is covered with a tough but flexible **cuticle** to which nematodes owe much of their success. It protects parasitic species from the host's immune response or digestive enzymes, and free-living species from adverse surroundings. One species of roundworm even thrives in vinegar. Under the body surface are longitudinal muscle fibres which enable roundworms to move by lashing from side to side or by eel-like undulations that sweep along the body from front to rear.

In contrast to flatworms, the sexes are separate and the reproductive system simple. During mating, sperm are transferred from the male to the female. Parasitic species can be extremely prolific: a female *Ascaris* may lay 200 000 eggs in a day.

Parasitic roundworms

Some nematodes are parasites of animals and plants. Most of these worms cause little or no harm (about 50 harmless species occur in humans), but others cause serious diseases, particularly in the tropics and subtropics.

Ascaris lumbicoides lives in the human small intestine. It feeds on the host's digested food which it sucks into its body through the pharynx. The adult worm, which may be 40 cm long, does not feed on the intestinal lining, but it may occur in such large numbers that the intestine becomes blocked. A person suffering from roundworm infection may have as many as 5000 worms. People become infected by eating food contaminated with the worm's eggs.

The hookworm *Ancylostoma*, of Africa and other warm parts of the world, is much smaller than *Ascaris* but more damaging. It rasps at the intestinal lining with three pairs of **hooked teeth**. Hookworm disease is characterised by severe anaemia and fatigue. In parts of Asia, 75% of the population are infected. The young worms live in the soil and infect humans by boring through the skin.

Some of the harm done by these and other parasitic roundworms is caused by the young worms which, having entered the host, wander through the body before maturing into adults in the intestine.

One of the most serious parasitic roundworms of plants is the eelworm, *Heterodera*, which attacks the roots of potatoes, tomatoes, cucumbers and sugar beet. The roots of a single potato plant may be infested with 40 000 eelworms.

Not all parasitic roundworms are a nuisance to humans. Some parasitise insect pests and are used as **biological control agents**. For example, a roundworm is used to control the vine weevil, which eats the roots and leaves of vines.

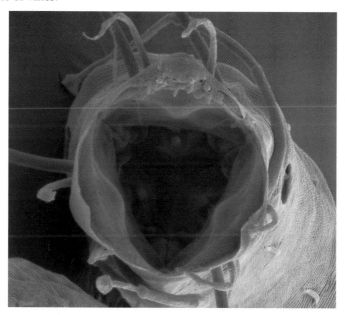

Figure 4.6 Scanning electron micrograph of the head end of a nematode with its mouth wide open (magnification x2000). This particular species is very small and lives in sea water between grains of sand. Through its feeding habits it helps to keep sandy beaches clear of bacteria and pollutants and is therefore important ecologically. Notice that the body is round in cross-section, a characteristic of nematodes.

4.4   Annelids (Phylum Annelida)

Annelids include earthworms, leeches and marine bristle worms. The annelid body plan is illustrated in figure 4.7. The most noticeable feature is the division of the body into a series of **segments** (see figure 4.7a). Many of the organs are repeated in most of the segments. This serial repetition of body parts is known as **metameric segmentation.**

Outwardly, metameric segmentation shows itself as a series of rings encircling the body (the word *annelid* means 'ringed'). Each ring consists of a fold of epidermis which tucks in to become continuous with a partition called a **septum** inside the body. The septa separate one segment from the next.

Annelids have a fluid-filled **body cavity** between the body wall and the gut (see figure 4.7b). This cavity is lined with mesoderm and is called a **coelom**. The coelom contains the organs.

A small **brain** on the dorsal side of the head is connected by a nerve ring to a **ventral nerve cord** from which **segmental nerves** pass to the body wall. This arrangement of the central nervous system, with a brain above the gut and nerve cord beneath it, is typical of annelids and related invertebrate phyla (see figure 4.7c).

Movement and locomotion

When an annelid moves, the muscle tissue in the body wall contracts against the fluid in the coelom, which is under pressure and serves as a **hydrostatic skeleton**. The septa between successive segments divide the coelom into a series of watertight compartments. This means that a change in pressure in one part of the worm is not immediately transmitted to other parts, so localised changes in shape can occur.

In an earthworm, these changes in shape take the form of bulges which pass along the body from front to rear (see figure 4.9). This method of loco-motion enables earthworms to burrow through soil. Where a bulge occurs, bristle-like **chaetae** are protruded from the body wall, and these enable the worm to grip the walls of its burrow.

Chaetae are a characteristic feature of annelids and the division of the phylum into classes is based on them. Oligochaetes, to which the earthworm belongs, have relatively few chaetae. Polychaetes, however, which include marine bristle worms such as the ragworm *Nereis*, have numerous chaetae which in some species are borne on paddle-like **parapodia**. And hirudineans (leeches) have no chaetae at all – they appear to have been lost in the course of evolution.

Gaseous exchange and transport

Most annelids lack special surfaces for gaseous exchange. Their long thin form gives them a large enough surface–volume ratio for gaseous exchange

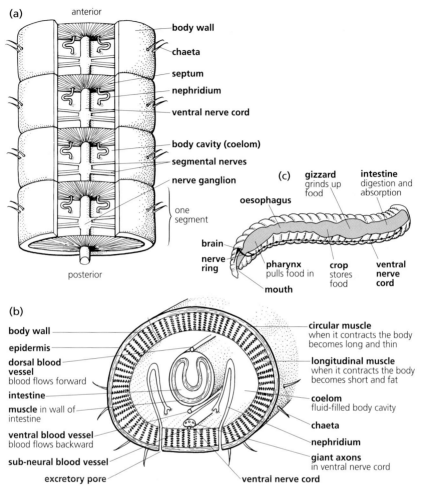

Figure 4.7 The annelid body plan is here illustrated by three views of an earthworm. (*a*) View from above showing metameric segmentation as seen through the body wall. (*b*) Transverse section (intestinal region). (*c*) Side view of the head region showing the gut and central nervous system. ((*c*) adapted from Keeton W. T. 1980 *Biological Science*, 3rd edition, Norton.)

to take place by diffusion through the moist skin. The only exceptions are certain polychaetes such as lugworms, which have feathery gills projecting from the body wall.

However, in contrast to flatworms, annelids have a **circulatory system**, which carries oxygen from the skin, and digestive products from the intestine, to all parts of the body. The blood is kept on the move by waves of contraction that sweep forward along the dorsal vessel. Some annelids have one or more simple **hearts**, which may help to propel the blood.

The blood contains a **blood pigment**. In the earthworm, this is iron-containing **haemoglobin**, but other pigments, such as **chlorocruorin**, are also found in annelids. It has been found that earthworm haemoglobin, like

Box 4.2 The coelom

The development of a cavity in the mesoderm – the coelom – was an important advance in the evolutionary history of animals. It is a fundamental feature of the body plan of most animals, including ourselves. Such animals are referred to as **coelomates**. Animals which lack a coelom – cnidarians and flatworms, for example – are called **acoelomates** (see figure 4.8).

Having a coelom carries a number of advantages. It houses many organs and enables the body wall and gut wall to move independently of each other.

In some coelomates, notably molluscs and arthropods, the coelom is reduced in size and the body cavity is a blood-filled haemocoel. However, in most coelomates, the coelom is large and is the main body cavity; the abdominal cavity of humans and other vertebrates, for example, is coelomic.

In annelids such as the earthworm, the coelom, full of fluid under pressure, has a hydrostatic function and is important in locomotion. In other coelomates, including humans, the coelom contains less fluid and its main role is to act as a lubricant allowing the organs to slide against each other without abrasion.

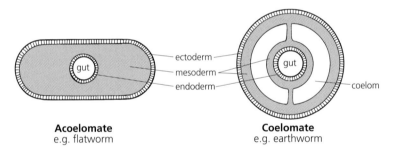

Acoelomate
e.g. flatworm

Coelomate
e.g. earthworm

Figure 4.8 An acoelomate and a coelomate compared in diagrammatic transverse sections.

mammalian haemoglobin, has a high affinity for oxygen. However, it is dissolved in the plasma rather than packed into red blood cells, and it is only a tenth as concentrated as in human blood. The oxygen-carrying capacity of the blood is therefore considerably less than that of mammals.

Excretion, osmoregulation and reproduction

Amongst the organs served by the circulatory system are the **nephridia**, of which there is a pair in nearly all the segments. Each nephridium is a coiled tube leading from the coelom to the exterior. It extracts nitrogenous waste from the blood and coelomic fluid, and controls the blood's ionic and

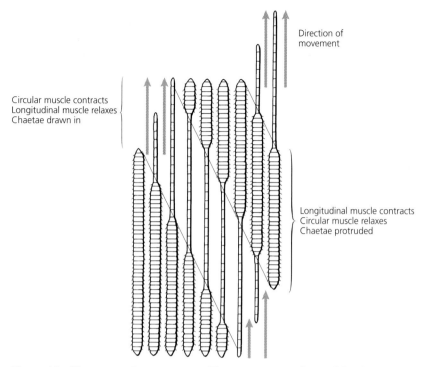

Direction of movement

Circular muscle contracts
Longitudinal muscle relaxes
Chaetae drawn in

Longitudinal muscle contracts
Circular muscle relaxes
Chaetae protruded

Figure 4.9 How an earthworm moves. The worm moves forward by the anterior end elongating and then forming a bulge which is propagated backward along the body. (Adapted from Kershaw D. R. 1988 *Animal diversity*, University Tutorial Press.)

osmotic concentration, removing unwanted substances and retaining wanted ones.

Marine annelids, such as *Nereis*, have a simple reproductive system. The sexes are separate but copulation does not occur. Instead, eggs and sperm are released into the surrounding sea water where fertilisation takes place externally. However, the earthworm, being terrestrial, has internal fertilisation. Earthworms are hermaphrodites and during mating, which takes place on the surface of the ground at night, sperm pass from each worm into the other.

Response to stimuli

Receptor cells sensitive to touch, chemicals and light are scattered all over the skin, but are particularly concentrated in the head region.

The main function of the brain is to receive impulses from the receptors in the head and transmit them, via the nerve cord, to the muscles. The brain plays little part in the coordination of locomotion. Locomotion is coordinated by the ventral nerve cord through a series of reflexes transmitted from segment to segment.

A striking feature of annelids is their **escape response**. If you prod the tentacles of a fanworm, for example, it withdraws into its tube; and if you touch the head of an earthworm lying on the surface of the soil, it will pull back quickly into its burrow. These responses are brought about by sudden contraction of the longitudinal muscle in the body wall. The contraction is initiated by impulses transmitted at high speed along greatly enlarged **giant axons** in the ventral nerve cord.

The rapid escape response of annelids contrasts with the much slower reactions of cnidarians and flatworms. It marks an important step forward in the evolution of survival mechanisms, and foreshadows the development of even more effective responses in arthropods and chordates.

Feeding in annelids

Annelids possess a variety of feeding methods, depending on the nature of their food.

Earthworms eat soil and leaf litter, which are pulled into the **pharynx** by muscular action and ground up in an expanded part of the gut called the **gizzard** (see figure 4.7c). The wall of the gizzard is muscular and its inner lining cuticularised, making it an effective pulverising organ. Digestion and absorption are completed in the intestine, whose surface area is increased by a longitudinal fold on the dorsal side. Indigestible matter is deposited from the anus on the surface of the ground as **worm casts**. As befits its burrowing life style, the earthworm's head is streamlined and devoid of awkward pro-truberances that might be damaged by the soil.

It is largely because of their feeding and burrowing habits that earth-worms are so useful to farmers and gardeners. They help to turn over, fertilise, aerate, drain and improve the texture of the soil. As much as 50 tonnes of worm casts may be deposited on the surface of one hectare of ground in a year. Through their activities, earthworms make ecological niches available to other organisms and so contribute to biodiversity. Charles Darwin wrote of them: 'It may be doubted whether there are many other animals which have played so important a part in the history of the world as these lowly organised creatures'.

The medicinal leech *Hirudo medicinalis* is an ectoparasite which sucks blood. Like the earthworm, it has a reduced head. The mouth is surrounded by a **sucker** with which the animal attaches itself to the host's skin. It then makes a small cut with three little **teeth**, and draws blood into its gut. An **anticoagulant**, produced by salivary glands, prevents the blood clotting. The gut has numerous branches in which the blood is stored, so one meal may last for weeks. Leeches are used on hospital patients for removing blood from swellings and for enhancing blood flow.

The free-swimming marine ragworm *Nereis* is carnivorous and has a well-developed head with an impressive array of sensory receptors. When feeding, it turns its pharynx outwards, exposing a pair of fang-like **jaws** with which it grasps its prey.

The peacock worm *Sabella* is sessile and lives in a protective **tube** made of sand grains stuck together with mucus. A 'fan' of delicate **tentacles** protrudes from the opening of the tube. The tentacles are ciliated and secrete mucus. A current of water is drawn through the fan by the cilia. Particles suspended in the water get caught up in the mucus and are drawn towards the mouth by more cilia. An ingenious sorting mechanism at the base of each tentacle ensures that only the smallest particles enter the mouth. These are the ones most likely to be of food value. Larger particles of sand or silt are either rejected or used for building the tube. Since this process involves sieving particles out of water, it is called **filter feeding**.

This brief survey of annelid feeding methods illustrates how members of a phylum may differ profoundly in their structure and occupy very different ecological niches. We can envisage that the various forms evolved by divergent evolution from a common ancestor. This is known as **adaptive radiation** and it is an important aspect of biodiversity.

4.5   Molluscs (Phylum Mollusca)

Snails, limpets, mussels and oysters are all molluscs. Their main characteristic is a hard **shell** into which they retreat in dry conditions and when danger threatens. Thus protected, there is no need for them to move rapidly, and in fact most of them are extremely slow or, in some cases, entirely sessile. And yet this phylum contains one group – squids and their relatives – which rivals fishes for speed and agility.

Most molluscs have, below the shell, a **head** which typically bears two pairs of short retractable tentacles, and a large fleshy **foot** which most molluscs use for locomotion. If you look at the diagram of the garden snail in figure 4.10, you will see that there is no clear demarcation between the head and foot – in fact, German textbooks refer to them together as the *Kopffuss* ('headfoot').

With one interesting exception, molluscs are unsegmented and the coelom is reduced. The body cavity consists of blood-filled spaces called the

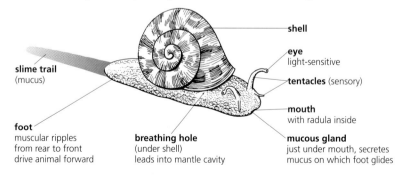

Figure 4.10   External features of the garden snail *Helix aspersa*.

haemocoel. The foot is protruded, and kept firm, by a hydraulic mechanism in which blood is pumped into its blood spaces. Nowhere is this better demonstrated than in the razor clam, which uses its long flexible foot for burrowing into mud.

The blood is kept flowing by a muscular **heart** which is surrounded by a **pericardial cavity**, a much reduced coelom. One or two **kidneys** link the coelom with the exterior, just as nephridia do in annelids. The main blood pigment of molluscs is **haemocyanin**, which is similar to haemoglobin but contains copper instead of iron. It is located in the plasma, not in blood cells.

In keeping with their sluggishness, most molluscs are herbivorous. On the lower side of the mouth cavity there is a flexible belt-like **radula**, covered with numerous small teeth, which moves backwards and forwards. This enables marine species, such as whelks, to rasp algae off rocks. Terrestrial snails and slugs have a hard ridge on the roof of the mouth cavity against which the radula can bite. This enables them to cut sizeable chunks out of leaves, as gardeners know to their cost.

Under the shell, in a position which varies from one group of molluscs to another, is a large space called the **mantle cavity**. In most molluscs, the mantle cavity contains one or more **gills**. However, the garden snail, being terrestrial, lacks gills and the mantle cavity functions as a **lung**.

The garden snail has an incredibly complex hermaphroditic reproductive system. However, most molluscs have separate sexes with simpler systems. In some groups, fertilisation is internal; in others it is external.

Adaptive radiation in molluscs

The molluscan body plan is based on three main components: the shell, the foot and the mantle cavity. In the course of evolution, these have become modified in various ways.

The ancestral mollusc is represented by a limpet-like creature called *Neopilina*, which was only discovered in the early 1950s. It has a conical shell and a mantle cavity surrounding a flat foot. The mantle cavity contains a series of gills and at the anterior end there is a small head. *Neopilina* is the only mollusc that shows metameric segmentation, suggesting a possible evolutionary connection with annelids.

A group of molluscs called chitons are rather like the ancestral form. However, the shell is flattened and divided into eight articulating transverse plates, which enable these animals to crawl over uneven surfaces, and roll up like woodlice.

Snails and other gastropods have a helical shell, twisted as a result of a process called **torsion** which occurs during development: the mantle cavity starts off at the posterior end and then rotates through 180 degrees so that it comes to lie above the head. The advantage is that the head can be pulled back into the mantle cavity for protection. However, the anus is brought forward too, with the unfortunate result that the animal defecates over its head.

Some gastropods have a greatly reduced shell. Such is the case with slugs and a marine group called nudibranchs. The foot of one type of nudibranch, the sea butterfly, is drawn out on either side to form a pair of 'wings' which flap up and down, enabling the animal to swim.

In mussels and their relatives (pelycopods), the shell is flattened and divided into two **valves** with a hinge along the dorsal side (in some classifications the pelycopods are called bivalves). The two valves are held together by powerful muscles. Large sheet-like gills hang in the mantle cavity on both sides of the foot, and these are used for filter-feeding. The head is greatly reduced and resides permanently inside the shell.

Most pelycopods are sessile, or move only very slowly. However, the scallop *Pecten* swims by opening and closing its valves very quickly. Its movements, though clumsy, help it to avoid predatory starfishes (see page 71). The shipworm *Teredo* uses its shell even more ingeniously: greatly reduced in size, it takes the form of a razor-sharp drill which bores into rock and timber. It can cause great damage to wooden jetties and ships' hulls.

But it is octopuses and squids that have departed most radically from the molluscan body plan. The shell is reduced to an internal strengthening 'pen' (or in octopuses lost altogether), freeing the body to become flexible and mobile. The mantle wall is thick and muscular, and the foot is incorporated into the head – hence the name 'cephalopod' (from Greek *kephale*, head, *podos*, foot) the class to which these molluscs belong. From the head projects a formidable array of **tentacles** armed with **suckers** for catching prey (see figure 4.11). The mouth cavity contains, in addition to a radula, a beak-like **jaw**. Below the mouth is a tubular **siphon**, also part of the foot, which connects the mantle cavity with the exterior. Some species are very large: the giant squid can be over 6 metres long (excluding the tentacles).

Figure 4.11 The cuttlefish, *Sepia officinalis*, a close relative of squids. It swims by undulations of the lateral fins, one of which is visible in the photograph. Notice the eyes and tentacles. The cuttlefish feeds on shrimps which it uncovers from the sand by blowing a jet of water at them. It catches them with its tentacles.

Octopuses and squids can move by jet propulsion. The mantle wall contracts powerfully, squirting a jet of water through the siphon. This propels the animal backwards, its streamlined posterior end minimising resistance. Its direction of movement is controlled by the siphon which can be bent in various directions. In flying squids, the jet is so powerful that the animal shoots out of the water and glides through the air, lift being provided by its fins and by membranes between the tentacles.

The jet propulsion mechanism is controlled by a set of **giant axons** which innervate the muscle tissue in the mantle wall. As in annelids, these axons transmit impulses rapidly and ensure that all the muscle fibres contract synchronously. In passing, it should be noted that our knowledge of how nerves transmit impulses derives in large part from experiments carried out on the giant axons of the squid. Invertebrates can, in fact, be immensely useful in unravelling fundamental biological problems – which is one good reason for studying them.

In keeping with their agility, octopuses and squids have well-developed image-forming **eyes** which, through convergent evolution, bear a striking resemblance to the vertebrate eye. The **brain** is large and remarkably well-developed; experiments with octopuses have shown that these animals have a learning capacity far greater than that of most other invertebrates.

Octopuses and squids are so unlike a typical mollusc that you might question whether they should be regarded as molluscs at all. However, a link is provided by *Nautilus*, a squid-like cephalopod with a coiled external shell. It is the only living derivative of an ancient group of molluscs known as ammonoids.

4.6   Echinoderms (Phylum Echinodermata)

Echinoderms include starfishes, sea urchins, brittle stars, feather stars, sea lilies and sea cucumbers. All are marine.

The most noticeable feature of echinoderms is their external covering of **calcareous plates** and **spines** from which the phylum gets its name (*echinoderm* means 'spiny-skinned').

Echinoderms are unique amongst animals in having a five-fold (**pentaradiate**) symmetry. This is clearly seen in the common starfish *Asterias*, which has five arms (see figure 4.12), but it is also apparent in other echinoderms such as sea urchins and sea cucumbers if you look hard enough.

A typical echinoderm, such as the starfish, has its mouth on the underside (the **oral surface**) in contact with the sea bed, and the anus on the top side (the **aboral surface**). However, some echinoderms have departed from this basic body plan. For example, sea cucumbers are elongated in the oral–aboral direction and lie on their side with the mouth at one end and the anus at the other. They have also lost their hard covering and become soft-bodied.

Characteristically, an echinoderm has, on its oral surface, hundreds of **tube feet** (see figure 4.12). Each tube foot can be extended or withdrawn and pointed in different directions. Starfishes and sea urchins can 'walk' on the sea bed by the concerted action of their tube feet. Collectively, the tube feet are amazingly powerful: *Asterias* uses them for feeding on bivalves, particularly mussels; the tube feet force the valves open.

Not all echinoderms rely on their tube feet for locomotion. Sea cucumbers, free from the constraints of a hard covering, move by contractions of the body wall like a worm, though they do use tube feet as well. In brittle stars, the arms contain articulating ossicles, like vertebrae, and are very flexible. These animals crawl by snake-like bending of their arms, the

Figure 4.12 The underside (oral side) of the common starfish, *Asterias rubens*, which lives in rockpools around British coasts. Here it is photographed through a sheet of glass to which it is attached by its tube feet.

tube feet playing little or no part in locomotion. And sea lilies are attached to the substratum by a stalk and use their tube feet for catching particles of food in a filter-feeding mechanism.

The tube feet are part of a system of water-filled tubes and sacs called the **water-vascular system**. Its function is to keep the tube feet fully charged with water. The term 'water-vascular' is misleading because it implies that this system has a circulatory function. However, echinoderms have no circulatory system. These animals are perfused with sea water, which provides their internal as well as their external environment. Osmoregulation and ionic regulation are therefore unnecessary, as is a circulatory system.

Gaseous exchange takes place by diffusion through the tube feet and through epidermal outgrowths from perforations in the external covering. Otherwise, there are no special surfaces for gaseous exchange, though here, too, sea cucumbers are an exception. They have a system of branched tubes, the **respiratory tree**, opening from – of all places – the posterior end of the gut. Contractions of this part of the gut pump water in and out of the respiratory tree and experiments have shown that the animal gets most of its oxygen this way.

In order to tackle the basic problems of life, echinoderms have developed solutions that are strikingly different from those of other phyla. They also hold a special interest for zoologists because of their possible evolutionary affinity with chordates, the phylum to which we belong.

Arthropods
(Phylum Arthropoda)

Lobsters, crabs, insects, spiders – these and many other animals are arthropods, one of the largest and most diverse phyla in the animal kingdom. It is so diverse that nowadays taxonomists are inclined to split arthropods into several separate phyla. Widespread in fresh water, the sea (including the ocean depths) and most land habitats, they display many adaptations for survival in hostile environments and have exploited almost all food sources, from wood to human blood.

5.1 The arthropod body plan

Arthropods are fundamentally similar to annelids, from which they probably evolved. Thus, they are bilaterally symmetrical and segmented, with a small brain linked to a ventral nerve cord. However, two features single them out from other invertebrates: they have **jointed appendages** and a hard protective **cuticle**, which can bend at the joints like articulated armour.

The appendages include **limbs**. Each limb consists of a series of rigid tubes lined with hard cuticle connected at the joints by a flexible membrane consisting of unhardened cuticle (see figure 5.1a). The limb is moved by muscles attached to the inside of the cuticle, so the cuticle functions as an **exoskeleton**.

The cuticle (see figure 5.1b) is composed of **chitin**, a tough nitrogen-containing polysaccharide. Chitin itself is soft but becomes impregnated with proteins, which are hardened (tanned) by quinones that form cross-linkages between the polypeptide chains. Only the outer part of the cuticle (the **exocuticle**) is hardened; the inner **endocuticle** remains soft and forms the flexible membrane at the joints. The surface of the cuticle is covered with a thin layer of material impregnated with wax (the **epicuticle**), which makes it waterproof.

Arthropods shed their cuticle at intervals and form a new one underneath, thus providing an opportunity for growth to take place. The periodic shedding of the cuticle is called **moulting** or **ecdysis** (see box 5.1). Arthropods are vulnerable to predators when they shed their cuticle, so they usually moult in a secluded place.

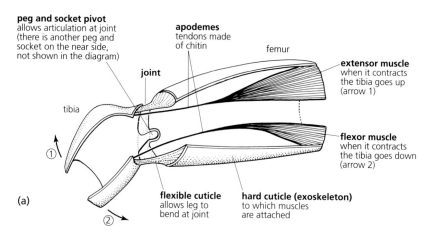

peg and socket pivot
allows articulation at joint
(there is another peg and
socket on the near side,
not shown in the diagram)

apodemes
tendons made
of chitin

femur

joint

extensor muscle
when it contracts
the tibia goes up
(arrow 1)

tibia

flexor muscle
when it contracts
the tibia goes down
(arrow 2)

(a)

flexible cuticle
allows leg to
bend at joint

hard cuticle (exoskeleton)
to which muscles
are attached

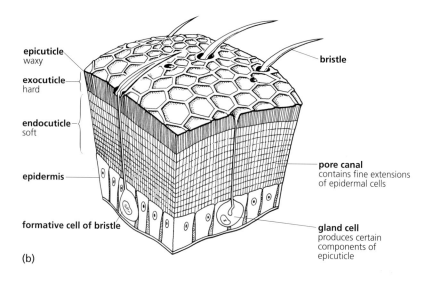

epicuticle
waxy

exocuticle
hard

endocuticle
soft

epidermis

formative cell of bristle

bristle

pore canal
contains fine extensions
of epidermal cells

gland cell
produces certain
components of
epicuticle

(b)

Figure 5.1 (*a*) The arthropod limb, based on the locust. Notice that the muscles are inside the exoskeleton. (*b*) The arthropod cuticle, based on an insect. The cuticle may be moulded into bristles, plates, pits or horns. Some bristles have receptor cells at the base that are sensitive to touch and vibration. (The limb is redrawn from Clarke W. M. & Richards M. M. 1976 *The locust as a typical insect*, Murray. The cuticle is redrawn from Kershaw D. R. 1988 *Animal diversity*, Chapman and Hall.)

Arthropods typically have a pair of **compound eyes** on the head. Instead of having a single lens and retina like our eyes do, the compound eye consists of hundreds – in some cases thousands – of closely packed units called **ommatidia**, each with its own lens and cluster of light-sensitive receptor cells (see figure 5.2).

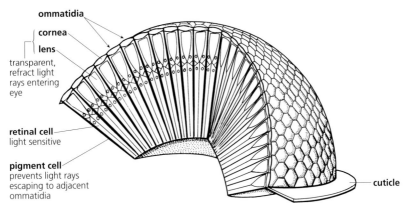

ommatidia

cornea

lens
transparent,
refract light
rays entering
eye

retinal cell
light sensitive

pigment cell
prevents light rays
escaping to adjacent
ommatidia

cuticle

Figure 5.2   The compound eye. Each ommatidium is a self-contained visual unit and sees things independently of the others. The image is composed of numerous dots, rather like a newspaper photograph seen through a magnifying glass. Most of the light refraction is achieved by the cornea, relatively little by the lens. The cornea is part of the cuticle and is shed every time the animal moults; the lens is gelatinous, secreted by surrounding cells. (Redrawn from von Frisch 1964 in Wigglesworth V. B. *The life of insects*, Weidenfeld and Nicolson.)

The compound eye can form an image, but its resolving power is poor compared with vertebrate and octopus eyes. However, in insects like the housefly, the compound eyes cover a large area of the head, giving a wide field of vision for spotting movements. In crustaceans, such as lobsters and crabs, the compound eyes are at the ends of mobile **eye stalks**.

Most arthropods also have, on the head, at least one pair of appendages called **antennae** ('feelers'), whose numerous joints enable them to bend in various directions. They usually carry receptors sensitive to taste, smell, touch and vibrations.

Close to the mouth are jointed appendages called **mouthparts**, which usually include a pair of jaw-like **mandibles** together with other structures for tasting and manipulating the food. In insects with specialised diets, the mouthparts are modified for dealing with particular types of food (see page 83).

In arthropods, the segments are not separated by septa, as they are in annelids, and there is no coelomic body cavity. Instead the body cavity is a blood-filled **haemocoel** and the coelom is greatly reduced. A muscular **heart** pumps blood into the haemocoel, either directly or via arteries; blood returns to the heart via small holes in the heart wall. In some arthropods an oxygen-carrying pigment (usually haemocyanin but sometimes haemoglobin) is present in the plasma.

Gaseous exchange occurs in several different ways. Aquatic arthropods such as lobsters and crabs, have **gills** attached to the bases of certain appendages. Spiders have **lung books**, pouches containing thin sheets like the leaves of a book; and some arthropods – notably insects – have a **tracheal system**, which is described in the next section.

Box 5.1 Moulting

Moulting (ecdysis) is best illustrated by insects, though the principles are the same in other arthropods too.

When an insect is ready to moult, the epidermis secretes a moulting fluid, which contains enzymes that dissolve the soft endocuticle, leaving only the hard exocuticle (see figure 5.3). Meanwhile, the epidermis produces a new cuticle under the old one. Substances in the waxy layer of the new cuticle prevent it being dissolved by the moulting fluid.

Then the insect expands (usually by swallowing air or water or by feeding) and the old exocuticle splits along lines of weakness and is cast off. The new cuticle, still soft, is stretched, allowing the animal to grow. The outer part of the new cuticle then hardens.

The pore canals play an important part in moulting. They carry the products of dissolution of the old cuticle back to the epidermis, and the hardening agents from the epidermis to the outer part of the new cuticle.

How is moulting controlled? As the time approaches for it to happen, neurosecretory cells in the brain – triggered by stimuli which vary in different insects – produce a hormone which flows into the thorax. When enough of this hormone accumulates, it stimulates a gland in the thorax to secrete a moulting hormone, which brings about the actual moult followed by growth.

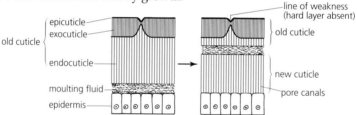

Figure 5.3 Two stages in the moulting of an insect. As quickly as the old cuticle is dissolved by the moulting fluid, a new cuticle is formed underneath.

5.2   Insects (Class Insecta)

Many insects are serious agricultural and garden pests, and vectors of plant and animal diseases. Others dazzle us by their beauty and are valuable for pollinating plants and in many other ways. They are important components of food chains in fresh water and on land where they link plants with carnivores such as birds and mammals. Many are scavengers, occurring in soil and leaf litter, where they help to initiate decay. Insects are superbly adapted to life on land, and this is reflected in most aspects of their biology.

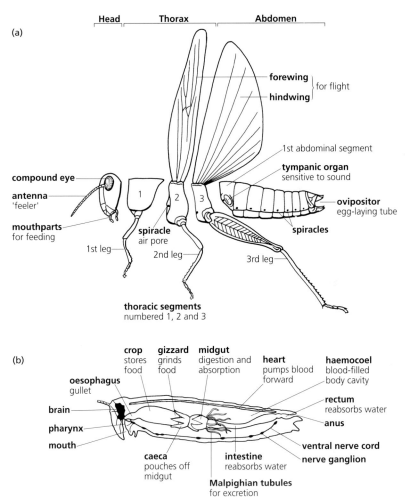

Figure 5.4 (*a*) The external and (*b*) internal anatomy of a typical insect, based on the locust. (Redrawn from Urquart F. A. 1965 *Introducing the insect*, Warne.)

The general anatomy of an insect is shown in figure 5.4. The body is divided into **head, thorax** and **abdomen**; all three thoracic segments bear **legs**, and the second and third bear **wings**. Internally, the gut and other organs lie in the blood-filled body cavity.

Gaseous exchange

The cuticle on either side of the thorax and abdomen is perforated by a series of small holes called **spiracles**, which lead to the **tracheal system** (see figure 5.5a). Oxygen and carbon dioxide move through the system by diffusion, aided in some insects (locusts, for example) by movements of the body wall. The spiracles may be guarded by **hairs** or **valves** which prevent excessive evaporation of water from inside the body (see figure 5.5b).

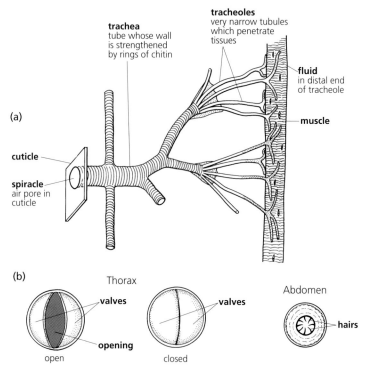

Figure 5.5 Tracheal system of an insect: (*a*) general view of the system, and (*b*) spiracles of the locust. The final diffusion of oxygen to the tissues takes place through the fluid in the distal ends of the tracheoles. When the muscles are active, this fluid is withdrawn by osmosis into the muscle tissue, thereby speeding up diffusion of oxygen in the tracheoles. Ventilation movements draw air through the anterior spiracles and expel it through the posterior ones. When a locust is resting, the valves of the thoracic spiracles close, preventing evaporative water loss. (Tracheal system redrawn from Meglitsh P. A. 1967 *Invertebrate zoology*, Oxford. Details of spiracles redrawn from Clarke A. M. & Richards M. M. 1976 *The locust as a typical insect*,

The tracheal system, relying on diffusion as the main means of transporting gases, is said to limit the body size and activity of insects. However, while insects certainly seem to be small (most are less than 2 cm long), they can be very active and apparently strong. The rhinoceros beetle *Xylorctes thestalus* can walk with a load 100 times its own weight – equivalent to a human carrying a double-decker bus. It manages to do this with a remarkably low metabolic rate, using five times less energy than would be expected – and it can keep this up for half an hour or more.

Since the tracheal system supplies oxygen direct to the tissues, the blood does not need to carry oxygen. Consequently, insects do not have a blood pigment and their blood is colourless. The only exceptions are the larvae of certain midges (*Chironomus* spp.) that live in stagnant, oxygen-deficient water. They possess haemoglobin and are known as 'blood worms' because of their red colour.

Water conservation

Insects' phenomenal powers of water conservation are due partly to their waterproof cuticle, but also to their method of excretion. This is carried out by the **Malpighian tubules**, a bunch of narrow, blindly ending tubes leading from the gut (see figure 5.4b). They extract soluble nitrogenous waste matter from the blood in the haemocoel and convert it into solid crystals of **uric acid**. In the intestine and rectum, water is reabsorbed from the excretory matter and faeces and taken back into the blood. Consequently, water that would otherwise be lost is retained in the body.

The ability of insects to conserve water enables them to survive in the driest of environments and has undoubtedly contributed to their success. A few species are even able to absorb water from vapour in the air.

Locomotion and response

Insects walk, run and in some cases jump using their legs. The musculo-skeletal basis of this is the same as in other arthropods.

Insects, however, are the only invertebrates that can fly. The **flight muscles** are not attached to the wings themselves but to the cuticle surrounding the thorax. By changing the shape of the thorax, the muscles make the wings move up and down (see figure 5.6). The mechanics of the system enables certain midges to beat their wings over 1000 times per second. Reflexes, initiated by receptors sensitive to airflow and distortion of the cuticle, help to maintain stability during flight.

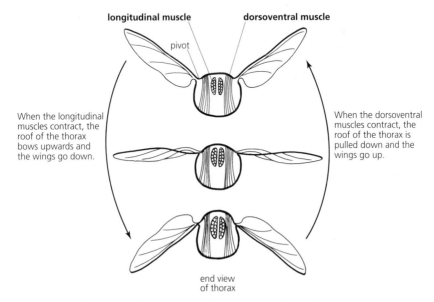

longitudinal muscle   dorsoventral muscle

pivot

When the longitudinal muscles contract, the roof of the thorax bows upwards and the wings go down.

When the dorsoventral muscles contract, the roof of the thorax is pulled down and the wings go up.

end view of thorax

Figure 5.6 The indirect flight muscles of an insect and how they operate the wings. (Redrawn from Wigglesworth V. B. 1964 *The life of insects*, Weidenfeld and Nicolson.)

Box 5.2 Evolving together: insects and flowering plants

Many species of flowering plants are pollinated by insects (see page 50). The insect picks up pollen from the anthers of one flower and deposits it on the stigma or stigmas of another flower, usually a different individual of the same species. The association between the flowering plant and the insect benefits both species: the insect obtains food from the plant (nectar and in some cases pollen too) and the plant gets its egg cells fertilised.

To entice insects to visit them, flowering plants have all sorts of adaptations such as brightly coloured petals, attractive scents and sugary nectar. Of course, what is attractive to an insect may not necessarily be attractive to us: the flowers of plants that attract flies smell, and in some cases look, like rotting flesh.

Some insects are attracted by a wide range of plant species, but plants lose out from such indiscriminate insects because their pollen may be taken to another species of plant with which it is incompatible. Other insects are attracted to *particular* plant species. This is more useful because the visitor will take pollen from, and deliver it to, plants of the same species.

Various adaptations help the insect to pollinate the plant. For example, some orchids look like bees and wasps – so much so that these insects try to mate with them. A certain species of orchid in Central America drugs its visitor with intoxicating nectar, causing it to stagger around the flower and get showered with pollen.

Sometimes the relationship between insect and flowering plant is highly specific. For example, yuccas are pollinated by a small moth which takes pollen from one plant, moulds it into a ball and sticks it on a stigma of another plant. The moth lays a few eggs inside the ovary and the young caterpillars feed on some of the developing ovules; the rest of the ovules form seeds in the usual way. By pollinating the plant, the moth ensures that the ovules develop. This is essential for the continuation of both the moth and the plant; indeed neither could exist without the other.

Such pollination mechanisms require the biology of the insect to correspond with that of the flowering plant. The relationship between the plant and its pollinator is the result of **coevolution**, the evolution of two species in response to each other.

Coevolution between flowering plants and insects is not confined to pollination. It has also produced insects that look like parts of plants such as flowers, thorns, leaves or twigs. Thus camouflaged, the insect is protected from attack by predators such as birds. An adaptation that prevents an organism being seen by predators is called **crypsis** and, although found in other groups of animals too, insects have refined it to perfection.

Anyone who has tried to swat a fly will know how quick an insect's responses can be. This is achieved by having sensitive receptors, nerves that transmit impulses quickly, rapidly contracting muscles and efficient coordination. In many insects the cuticle is thin, which helps to lighten the body and make the animal more agile.

The sexes are separate, the reproductive organs being located at the posterior end of the abdomen. The male has a 'penis' through which, during mating, sperm are deposited in a **sperm receptacle** inside the female. The eggs are fertilised, one by one, as they pass the opening of the sperm receptacle. The fertilised eggs, replete with a food reserve of nutritious yolk and encapsulated within a protective shell of tanned protein and wax, are then laid – often in vast numbers – through the **ovipositor** (see figure 5.4a).

Males and females find each other mainly by smell or sound. Sounds are often produced by rubbing parts of the cuticle together, a process called **stridulation**. Grasshoppers, for example, rub the serrated inner surface of the upper part of the hind legs against the hardened edges of the wings to produce the familiar 'buzzing' sound. In the mole cricket *Gryllotalpa gryllotalpa*, the sound is amplified by the shape of its burrow, which acts like a loudspeaker. The sounds are registered by **tympanal organs** on either side of the abdomen; they are like ear drums and vibrate in sympathy with the sound waves.

Sexual attraction by smell involves the release of volatile **pheromones**. A pheromone is a substance produced by one individual which influences the behaviour, or development, of another. Sex pheromones are particularly important in nocturnal insects such as moths. The pheromone, produced by the female, is detected by receptors on the male's antennae, which may be enlarged or feathery to increase their surface area. In some species, the receptors are so sensitive that a male may detect a female downwind ten kilometres away.

Insect life cycles

The life cycles of insects show **metamorphosis**. Metamorphosis is the profound, often rather sudden, change from the juvenile stage to the adult in the life cycles of certain animals including insects and amphibians (see page 96).

Two kinds of metamorphosis occur in insects. In **complete metamorphosis**, the egg develops into a **larva**, which differs from the adult in its structure, feeding habits, behaviour and the fact that it cannot reproduce. It feeds and grows, usually moulting several times, then becomes a dormant **pupa** surrounded by a protective case. Inside the pupa, small groups of embryonic cells called **imaginal buds** develop into the adult organs, the necessary nutrients coming from the larval tissues, which are broken down by phagocytes into a fluid mass. When the adult body is complete, and environmental factors permitting, the insect emerges from the pupa as the adult (**imago**), which reproduces sexually, often with great haste.

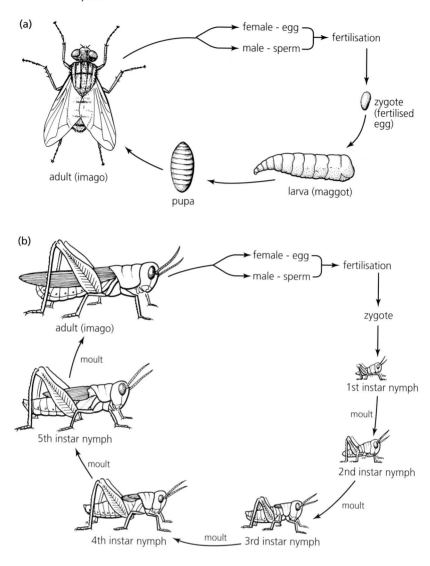

Figure 5.7 The life cycles of insects. (*a*) The housefly has complete metamorphosis. During the summer the females lay their eggs on decaying meat, vegetable matter or faeces. The eggs hatch into larvae (maggots) which burrow into, and feed on, the decaying material. They then leave their food and pupate. After a few weeks, the adult flies emerge from the pupae. Pupae formed in the autumn can survive the winter and give rise to adult flies the following spring. (*b*) The locust has incomplete metamorphosis. The type of locust on which these diagrams are based inhabits desert areas. The female lays her eggs in the sand. The eggs hatch after about two weeks and the young nymphs crawl out of the sand. They moult five times, growing a bit each time. As they have no wings yet, they cannot fly. They can only walk and hop. They feed on plants, using their mandibles for chewing. After about two months, the final moult takes place, the wings expand and the adult is formed. The adults, too, feed on plants and can do immense damage to crops.

Insects with this sort of life cycle include butterflies and moths (in which the larva is the caterpillar) and flies (in which the larval stage is the maggot). They are called **holometabolous insects** (see figure 5.7a).

In **incomplete metamorphosis**, the egg develops into a **nymph**, which resembles the adult except that it is smaller, lacks wings and cannot reproduce. It feeds on similar food to the adult and moults several times, growing a bit each time. The series of nymphal stages are called **instars**. At the final moult, small wing buds on either side of the thorax expand and the sexually active adult is formed.

This sort of life cycle is shown by locusts, cockroaches and many other insects. They are called **hemimetabolous insects** (see figure 5.7b).

Metamorphosis permits a division of labour between a feeding juvenile stage and a sexually reproducing adult. In certain species of moth, for example, the larva has well-developed mouthparts and feeds voraciously, whereas the adult has reduced mouthparts and does not feed at all. The adult may live for only 24 hours, during which its sole function is to find a mate and reproduce.

Classification and diversity of insects

Insects are split into 29 orders, which can be grouped together on the basis of the presence or absence of wings and type of metamorphosis (see figure 5.8).

Insects are incredibly diverse, especially in their mouthparts and feeding habits. The story starts with the mouthparts of insects like the cockroach and locust, which chew their food (see figure 5.9a). In the course of evolution, the components of this basic set of mouthparts have been modified into long flexible tongues for probing flowers and sucking up nectar (bees and butterflies), sponge-like pads for dissolving solid food and sucking up the resulting fluid (houseflies) and hypodermic needles for piercing and sucking blood or plant juices (mosquitoes and aphids, see figure 5.9b). The diversity of insect mouthparts is a classic example of adaptive radiation.

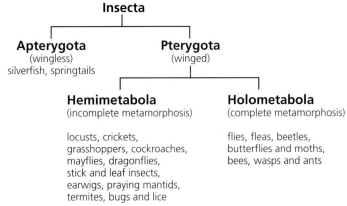

Figure 5.8 Classification of insects.

It is mainly because of their feeding habits that insects are so important for humans. Some are helpful (e.g. pollinators and those that feed on harmful species); others are a nuisance. Aphids, for example, insert their proboscis into the phloem of plants and feed on sugars and other nutrients flowing along the sieve tubes. Hundreds of aphids may feed on a single leaf. As well as damaging the plant's tissues, they transmit diseases: over 60% of plant viral diseases are transmitted by aphids.

Social insects

Social insects live in **colonies**; they include bees, wasps, ants and termites. The colony consists of several different types of individuals called **castes**, each of which does certain jobs. Here we look at termites.

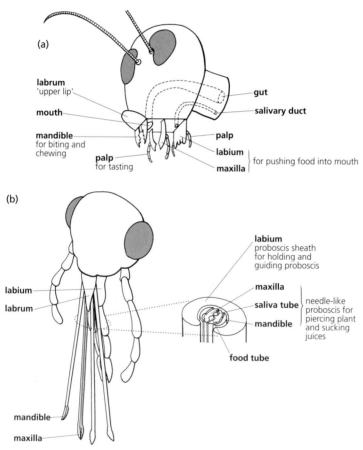

Figure 5.9 Mouthparts of (*a*) a locust and (*b*) an aphid. Notice that in these two insects equivalent (i.e. homologous) parts are differently constructed and perform different functions. Thus the locust has serrated mandibles for cutting and chewing whereas in aphids the mandibles form the sides of the needle-like proboscis through which plant juices are sucked. (Aphid mouthparts adapted from Kershaw D. R. 1988 *Animal diversity*, University Tutorial Press.)

The termite colony is started by a pair of winged individuals (male and female) which mate, lose their wings and become **king** and **queen** of the colony. Each day the queen lays thousands of eggs, which give rise to sterile wingless **workers** and **soldiers**. The workers build the nest, which consists of a labyrinth of interconnected chambers in tree trunks or dry timber, or underground. Ground termites construct their nests out of mud made by mixing soil with saliva. The chambers may be extensive and have tall mounds above the ground. As well as building the nest, the workers feed the queen, collect the eggs and rear the young (see figure 5.10). The soldiers defend the nest; they have powerful biting mandibles and, in some species, a gland on the head which squirts a repellent liquid at intruders.

When the colony is a few years old, fertile winged and wingless males and females are produced by the queen. The wingless individuals reproduce within the nest in which they were born. The winged ones fly away and mate, usually with termites from another nest, and start new colonies. At its most populous, a colony may contain three million individuals.

What causes the different castes to develop? Research has shown that a pheromone, secreted by the queen, is spread through the colony by the workers with their mouths. This pheromone prevents the young undergoing their final moult into adults, so they remain wingless and sterile and become workers or soldiers.

Termites feed on plant material brought to the nest by the workers. Some species feed only on wood and can cause severe damage to houses and other wooden structures, particularly in the tropics. To digest wood, the

Figure 5.10 Termite workers tending the queen in a termite nest. The queen has a greatly enlarged abdomen (up to 10 cm long in some species) and enormous ovaries which may produce 8000 eggs per day. The photograph shows the reproductive opening at the posterior end of the abdomen.

enzyme cellulase is needed. No known animal produces this enzyme. However, termites have cellulase-producing flagellates in their guts. These flagellates ingest pieces of wood swallowed by the termites and break them down into soluble sugars. The association between the termite and the flagellates is an example of **mutualism** (see page 14). The termite receives sugars from the flagellates, and the flagellates receive food and shelter from the termite.

Some species of termites collect fungi, which they cultivate on a bed of chewed vegetable matter in special 'fungus gardens' inside the nest. The fungus is fed by the workers to the king, queen and developing nymphs.

For a colony of social insects to work in a coordinated manner, its members must be able to communicate with each other. This is achieved by pheromones and also by behaviour. Worker bees, for example, can inform each other of the whereabouts of food sources by performing 'dances' and possibly by making sounds. Such methods of communication, though elaborate, appear to be completely instinctive, as indeed is most insect behaviour. Learning plays little part in the lives of most insects.

5.3   Other arthropods

Arthropods can be divided into a series of classes distinguished mainly by the number of legs.

Crustaceans (Class Crustacea)

Crabs, lobsters, crayfish, prawns, shrimps, barnacles, waterfleas and woodlice are all members of this diverse group. Most are aquatic, woodlice being the only truly terrestrial ones and even they need a damp environment to survive (see box 1.2 on pages 10 and 11). Crustaceans are important components of marine and freshwater food chains, smaller species and larvae forming part of the plankton.

A typical crustacean has **two pairs of antennae** on the head, and other appendages – including a variable number of legs – occur along the length of the body. The appendages are usually branched, and some of them bear **gills** for gaseous exchange.

The segmented body is typically divided into a **cephalothorax** (head and thorax combined) and **abdomen**. The cephalothorax is usually covered by a protective 'shield' called the **carapace**. The exoskeleton is made extra hard with calcium salts.

In lobsters and prawns, the abdomen contains a pair of thick longitudinal muscles, well known to gourmets of seafood. When these muscles contract, the abdomen flips forward and the animal shoots backward, an effective escape reaction mediated by giant axons (see page 70).

Some crustaceans depart radically from the basic arthropod body plan.

For example, barnacles are attached to the substratum by the head and use their limbs for filter feeding. Even more bizarre is a relative of barnacles called *Sacculina*, which parasitises crabs. The larva invades the host and fills it with a hideous mass of branching fibres through which the main body of the parasite (a tumour-like protuberance on the lower side of the host's abdomen) receives nourishment.

Centipedes and millipedes (Classes Chilopoda and Diplopoda)

The body of these creatures is elongated with a pair of legs on nearly every segment. Most centipedes have between 15 and 30 pairs of legs, though some species have more. Centipedes are predators: prey is seized by a pair of **poison claws**, then chewed by the mandibles. The giant Brazilian centipede *Scolopendra subspinipes* (see figure 5.11) may reach a length of 25 cm and its poison can cause excruciating pain.

Millipedes are herbivorous and some feed on the roots of crop plants. Successive pairs of segments are fused, so each segment appears to bear two pairs of legs. The number of legs is usually greater than that of centipedes; one species – *Illacme plenipes* of California – has over 350 pairs.

At one time, centipedes and millipedes were placed together in a single class called Myriapoda (which literally means 'thousands of legs'). However, detailed comparison of their structure shows them to be less closely related than they first seemed, so now they are placed in separate classes.

Arachnids (Class Arachnida)

Spiders, harvestmen, scorpions, mites, ticks and horseshoe crabs all belong to this class, characterised by having **four pairs of legs**. Arachnids lack compound eyes, antennae and mandibles. The role of the mandibles is performed by a pair of claw-like **chelicerae**, which are mainly used for grasping food, and, behind these, a pair of **pedipalps** for manipulating and

Figure 5.11 The giant centipede *Scolopendra* of Central America.

crushing it. In spiders, the chelicerae are fangs through which poison is injected into the prey. The poison of the black widow spider can be fatal, but that of the large and much feared hairy tarantula is no worse than a wasp sting.

Spiders are best known for their **silk**, which they use for spinning webs, lining burrows, catching prey, protecting their eggs, preventing themselves falling, sailing through the air ('ballooning') and for various aspects of courtship and mating. Silk is secreted as a liquid protein from glands at the posterior end of the abdomen. As the protein leaves the body it passes through nozzle-like **spinnerets** in which it hardens.

Harvestmen are like spiders but have exceptionally long and slender legs. They also possess, close to the base of the first pair of legs, a pair of glands which secrete a fluid that is distasteful to predators.

Scorpions have powerful claws at the end of much enlarged pedipalps, and the elongated abdomen has a curved spine at the posterior end from which a poison (not usually fatal to humans) can be injected into prey. The ancestors of scorpions terrorised the Silurian seas 450 million years ago: one species was over three metres long!

Mites and ticks include predators, scavengers and ectoparasites. The predators and scavengers have jaw-like chelicerae, but the parasitic forms have feeding structures adapted for sucking blood or plant juices. Some are serious pests of humans, farm animals and crops, damaging tissues and transmitting diseases. Scabies is caused by a tiny mite that burrows into human skin and causes intense itching, and the faeces of house mites breathed in with dust can cause asthma.

Horseshoe crabs are large marine arachnids that are thought to have changed very little during their evolutionary history. They occur in large numbers on the east coast of North America and are the only arachnids to have compound eyes.

Classification of arthropods

The conventional division of the arthropods into five classes (crustaceans, chilopods, diplopods, insects and arachnids) is considered unsound by many taxonomists. They prefer to split them into three subphyla: **Crustacea** (branched appendages), **Uniramia** (unbranched appendages, including chilopods, diplopods and insects) and **Chelicerata** (chelicerae instead of mandibles, including arachnids). Some authorities go further and regard these three groups as separate phyla because they seem to have had independent evolutionary origins. In such a classification, the phylum Arthropoda does not exist at all!

Chordates (Phylum Chordata)

Most chordates have a vertebral column (backbone) and are therefore vertebrates; they include fishes, amphibians, reptiles, birds and mammals. In addition, there are a few invertebrate chordates, including the fish-like *Amphioxus*, the worm-like *Balanoglossus* (acorn worm) and sessile sea squirts. These invertebrate chordates, and echinoderms to which they are related, are living representatives of an ancient group from which the vertebrates are believed to have evolved.

6.1 The chordate characters

All chordates share six diagnostic features, known as the **chordate characters**, which are evident at some stage during their life history (see figure 6.1).

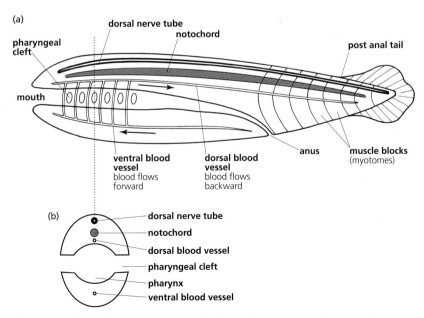

Figure 6.1 The body plan of a generalised chordate, showing the chordate characters: (*a*) viewed from the side, and (*b*) in transverse section.

1 A rod-like **notochord** runs more or less the full length of the body on the dorsal side. It consists of tightly packed cells, filled with gelatinous fluid, surrounded by a fibrous sheath which makes it tough but flexible. Its function is to support the body.

2 The central nervous system consists of a **dorsal tubular nerve cord**, which lies just above the notochord.

3 On either side of the body there is a series of blocks of muscle called **myotomes**. These, and the nerves that supply them, follow a repeated pattern along the body, which therefore shows metameric segmentation.

4 The pharynx is connected to the exterior on both sides of the body by a series of **pharyngeal clefts**. In invertebrate chordates they are mainly used for filter feeding, but in fishes they house the gills, which are used for gaseous exchange.

5 The anus is not at the extreme posterior end of the body but further forward, leaving a **post-anal tail** of variable length. It contains myotomes and in many chordates is the most obviously segmented part of the body.

6 Blood flows forward in a ventral vessel and backward in a dorsal vessel.

Chordates contrast sharply with other animal phyla such as annelids and arthropods in which notochord, pharyngeal clefts and post-anal tail are absent, the nerve cord is ventral and solid, and blood flows forward dorsally and backward ventrally.

Vertebrates

Vertebrates have the chordate characters plus a number of additional features. The notochord is present in the early embryo but is replaced later by the **vertebral column**. The dorsal nerve tube expands anteriorly to form the **brain**, while the rest of it becomes the **spinal cord**. Associated with the brain are **special sense organs** such as eyes and ears. Part of the ventral blood vessel expands to form the **heart**, which pumps the blood. The blood has haemoglobin in **red blood cells**, and several types of **white blood cells** form the basis of a highly developed **immune system**.

6.2 Fishes

'Fish' is an informal (non-systematic) term covering all aquatic vertebrates with fins and gills. They originated about 400 million years ago as heavily armoured jawless forms from which modern jawed fishes evolved. Today the only jawless fishes are lampreys and hagfishes, which have round suctorial mouths with a rasping tongue-like device and live ectoparasitically on other fishes, feeding on their tissues.

Jawed fishes fall into two main classes: **Chondrichthyes**, whose skeleton is made of cartilage, and **Osteichthyes**, whose skeleton is made of bone.

Bony fishes (Class Osteichthyes)

Over 20 000 species of bony fish occur in fresh water and the sea. They range in size from tropical gobies less than 1cm long to the ocean sunfish, which can be over 3m long and weigh more than 900 kg.

The external features of a typical bony fish are shown in figure 6.2. The **fins**, some paired and others unpaired, are supported by slender **rays** made of bone. The skin is covered with smooth flat **scales** which make it impervious to water; the **gills** on each side of the body are covered by a bony flap, the **operculum**. The **lateral line** is a narrow tube, open to the exterior, which contains receptors sensitive to pressure changes (vibrations) in the surrounding water.

All fishes have a **single circulation**: blood flows once through the heart for every complete circuit of the body, an arrangement which differs from that of mammals. The blood passes through two capillary systems (those of the gills and then the rest of the body) before it returns to the heart. Having been through two sets of capillaries, the blood is at a much reduced pressure, so the bloodflow in the veins tends to be sluggish.

Bony fishes possess a **swim bladder**, a gas-filled sac in the upper part of the body cavity, which provides buoyancy. There are two types of swim bladder, the **open type** and the **closed type**. The open type is connected by a short tube to the pharynx and contains atmospheric air, whereas the closed type has no connection with the pharynx. A fish with the open type swims to the surface of the water and gulps air to increase its buoyancy, and 'blows bubbles' to decrease its buoyancy. In a fish with the closed type, gas is secreted into, or absorbed from, the swim bladder by vascularised glands in its wall; such a fish can rise or fall in the water like a submarine.

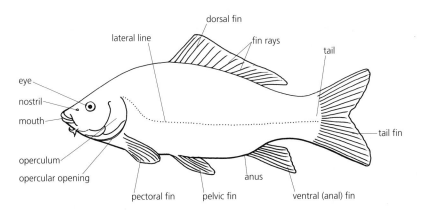

Figure 6.2   External features of a carp.

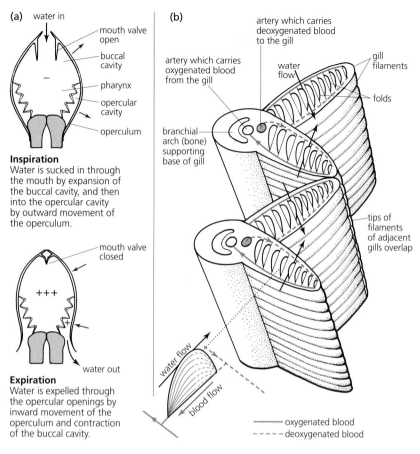

(a) water in

— mouth valve open

— buccal cavity

— pharynx

— opercular cavity

— operculum

Inspiration
Water is sucked in through the mouth by expansion of the buccal cavity, and then into the opercular cavity by outward movement of the operculum.

— mouth valve closed

+++

+

water out

Expiration
Water is expelled through the opercular openings by inward movement of the operculum and contraction of the buccal cavity.

(b)

artery which carries deoxygenated blood to the gill

artery which carries oxygenated blood from the gill

water flow

gill filaments

folds

branchial arch (bone) supporting base of gill

tips of filaments of adjacent gills overlap

water flow

blood flow

——— oxygenated blood
- - - - deoxygenated blood

Figure 6.3 Gaseous exchange in a bony fish such as a carp. (*a*) Horizontal sections of the head showing the gills and ventilation mechanism. + and − indicate the pressure of water in the buccal and opercular cavities relative to the outside water. The movements of the buccal cavity and operculum are timed so that a continuous stream of water flows over the gills. (*b*) The gills in detail. Notice that the tips of the gill filaments overlap, which means that all the water from the buccal cavity must flow over the gills before it leaves the body. The water flows in the opposite direction to the blood (counterflow system), ensuring that the blood takes up as much oxygen as possible before it leaves the gill. (The horizontal sections are redrawn from Hughes & Morgan 1973 in *Biol. Rev.* **48**: 419-75. The gill details are redrawn from Hughes G. M. 1974 *Comparative physiology of vertebrate respiration*, Heinemann.)

The gills are located on both sides of the buccal cavity. Water, drawn in through the mouth, flows over the gills and is expelled through the opercular openings (see figure 6.3a). The gills consist of stacks of folded, well vascularised **filaments**, which together present a large surface area for gaseous exchange between the blood and water (see figure 6.3b). The gills are orientated so as to maximise oxygen uptake, and the diffusion distance between the water and the blood is very short (about 0.5 μm).

Most bony fishes swim by sweeping the tail from side to side, the propulsive effect enhanced by the large surface area of the tail fin (see figure 6.4). The highly mobile pectoral and pelvic fins provide lift on the principle of a hydrofoil, an underwater 'wing' which lifts a boat's hull out of water at high speeds. They are also used for steering, braking and stability. Some species are extremely agile.

In eels, lateral undulations pass from front to rear, generating a forward force all along the body. This method of propulsion is also effective on land and enables eels to travel across country.

Any aquatic animal with permeable surfaces such as gills may gain or lose water by osmosis. In freshwater bony fishes, the solute concentration of the blood and tissue fluids is higher than that of the surrounding water, so water enters the body by osmosis, thus diluting the body fluids. This is remedied by eliminating surplus water through the kidneys and by actively taking up salts from the surrounding water through special cells in the gills. Nitrogenous waste is excreted as ammonia which, though highly toxic, is greatly diluted by the copious flow of water from the kidneys.

In marine bony fishes, the solute concentration of the blood and tissue fluids is lower than that of the surrounding water, so water *leaves* the body

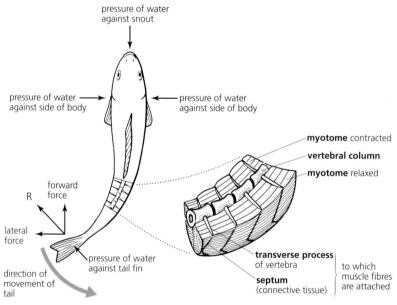

Figure 6.4   Locomotion in a bony fish such as a carp. As the tail sweeps from left to right, the pressure of water against the tail fin produces a force R on the fish which can be resolved into two components: a forward force which drives the fish forward and a lateral force which tends to deflect the fish sideways. Any tendency for the fish to swing sideways as the tail lashes back and forth is counteracted by the pressure of water against the flattened sides of the body and dorsal fin, and any resistance to forward movement caused by the pressure of water on the snout is counteracted by the streamlined shape of the fish.

by osmosis, thus concentrating the body fluids. This is remedied by the kidneys reabsorbing water and by the cells in the gills actively expelling surplus salt. Nitrogenous waste is excreted as urea, which is less toxic than ammonia and does not require as much water for its elimination, and as **trimethylamine oxide**, which is totally non-toxic.

Most bony fishes have **external fertilisation** and show little care of the eggs or young. To overcome the slender chances of survival, vast numbers of eggs are produced: as many as 7 million in a single spawning by a female cod, and up to 30 million by the massive sunfish *Mola mola*. But there are exceptions. For example, guppies have internal fertilisation and give birth to live young (**viviparity**), and male sticklebacks guard their fertilised eggs and aerate them by fanning them with the tail. Species which look after their young produce fewer eggs. Certain cichlids, for example, produce a small clutch of eggs, which they keep in their mouths; after the eggs have hatched the baby fish (fry) pop back into the parent's mouth if danger threatens.

Bony fishes show many specialised adaptations. For example, the mudskipper *Periophthalmus* has limb-like pectoral fins with which it crawls over the mud and even climbs trees; the electric eel *Gymnarchus* has modified muscles which, instead of contracting, produce electric shocks of over 500 volts for stunning prey; and the flying fish *Exocoetus* propels itself out of the water with its tail and then glides through the air for several hundred metres on wing-like pectoral fins.

Bony fishes that live in oxygen-deficient water can breathe air. Some use the swim bladder for gaseous exchange; others use various parts of the gut including the stomach, intestine and even the rectum; and some have developed lungs.

Cartilaginous fishes (Class Chondrichthyes)

Sharks and rays belong to this class. With the exception of the South American sawfish *Pristis*, which swims up estuaries, they are all marine. As well as having a cartilaginous skeleton, the body tends to be flattened dorso-ventrally (very much so in rays) and the mouth is on the ventral side of the head rather than terminal. The scales (called **placoid scales**) are usually pointed and the fins are fleshy.

Cartilaginous fishes have no swim bladder and most species sink if they stop swimming, which is why many species spend most of their time on or close to the sea bed. In a shark's tail, there is more fin below the vertebral column than above it with the result that, as it sweeps from side to side, it keeps the hind end up in the water. The front end is held up by the large flap-like pectoral fins, which act as hydrofoils.

There is no operculum, and the gills open to the exterior on both sides of the body by five separate **gill slits**. Water is drawn in to the pharynx through the mouth and a pair of small holes called **spiracles**, and is then forced over the gills and out through the gill slits.

Like marine bony fishes, cartilaginous fishes face the potential problem of losing water by osmosis. However, urea is retained by the kidney instead of being excreted, and this raises the solute concentration of the blood and tissue fluids slightly above that of the surrounding water. Any surplus water that enters the body by osmosis is eliminated by the kidneys.

Fertilisation is internal: the male has a pair of **claspers**, formed from the inner side of the pelvic fins, which transfer sperm to the female during mating. Each fertilised egg is protected by a horny case in which the embryo develops. Nourishment is provided by a **yolk sac** attached to the underside of the embryo. Some of the larger sharks are viviparous, bearing live young.

Cartilaginous fishes include some notorious predators. One infamous eighteenth-century shark is reputed to have been caught with an entire suit of armour in its stomach!

6.3   Amphibians (Class Amphibia)

Amphibians include salamanders, newts, frogs and toads. They were the first vertebrates to invade dry land and this is reflected in many aspects of their biology.

The paired fins, characteristic of fishes, are replaced by **limbs**, which support and move the body on land. Typically, the limbs splay out in a 'press-ups' position, so energy has to be expended to hold the body off the ground.

Opening from the pharynx in adult amphibians is a pair of **lungs** for gaseous exchange; these are simple sacs, slightly folded internally to increase their surface area. Gulping movements of the floor of the buccal cavity and pharynx draw air in through a pair of **nostrils** and force it into the lungs. Gaseous exchange also takes place across the skin, which is permeable, well vascularised and kept moist by mucus secreted from glands beneath the epidermis.

In contrast to fishes, amphibians have a **double circulation**: after being pumped from the heart to the lungs (and skin), the blood returns to the heart from which it is pumped to the rest of the body. After passing through the capillary systems, the blood is at a higher pressure than in fishes and returns to the heart more quickly.

Ideally, a double circulation requires the heart to be divided into two halves, one for pumping deoxygenated blood to the lungs, the other for pumping oxygenated blood to the body. In amphibians only the atrium is divided, the ventricle remaining as a single chamber. This results in a certain amount of mixing of oxygenated and deoxygenated blood before it leaves the heart.

Amphibians have never completely forsaken their ancestral aquatic home. They return to water to breed. In frogs and toads the male clings to the female, but fertilisation is external and the eggs, though surrounded by

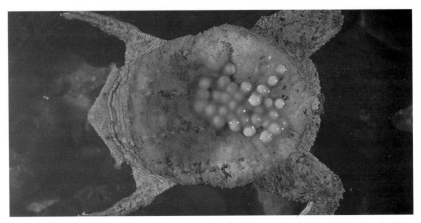

Figure 6.5   The Surinam toad *Pipa pipa* with eggs on its back. The eggs fit into depressions in the skin.

protective jelly, lack a shell. Each egg, if it survives, hatches into a larva – the **tadpole** – which is completely aquatic and breathes by means of gills. These are feathery and external to begin with, but later become internal (fish-like).

After a period of growth and development, the larva undergoes **metamorphosis** into the adult, which leaves the water and moves onto land. But most adult amphibians, with their moist permeable skin, are poorly adapted for terrestrial life and are confined to damp areas. In fact newts, though they can walk on land, prefer to live in ponds, where they propel themselves with their fish-like tail.

Some amphibians show intriguing adaptations for reducing their dependence on water. Certain tropical tree frogs lay their eggs in rain-filled plants or in hollows in trees, and foam-nesters lay theirs in watery mucus which they whip up into a froth with their hind legs.

Most amphibians show little or no care of their eggs and young, but there are exceptions. The midwife toad *Alytes obstetricans*, of central Europe, mates on land: the female extrudes long strings of eggs embedded in jelly, and the male, having fertilised them, wraps them round his hind legs and carries them until they hatch. The South American frog *Pipa* carries its eggs in pits on its back (see figure 6.5).

6.4   Reptiles (Class Reptilia)

Reptiles include lizards, crocodiles, alligators, snakes, tortoises and turtles. The skin is covered with protective **scales**, which are made of the protein **keratin** and make the skin dry and impermeable to water and gases. Most reptiles have two pairs of limbs which, like those of amphibians, splay out, so energy has to be expended in keeping the abdomen off the ground. Snakes, of course, have lost their limbs and move on land (or in some cases swim) by throwing their elongated body into lateral undulations.

Being impermeable, the skin cannot be used for gaseous exchange. This function is carried out entirely by **lungs**, which are more folded and therefore have a larger surface area, than those of amphibians. Air reaches the lungs via a **trachea** and **bronchi**, and the lungs are ventilated by in and out movements of **ribs**, a more efficient mechanism than the gulping of amphibians.

The circulation is double and the ventricle is partially divided into right and left halves for separating deoxygenated and oxygenated blood.

Reptiles are well adapted to living on dry land; in fact many species thrive in the desert. They conserve water by their impermeable skin and by excreting nitrogenous waste as solid **uric acid**. In this respect they are similar to insects (see page 79).

Reptiles control their body temperature behaviourally, warming themselves by basking in the sun and cooling off by burrowing or moving into the shade. Some reptiles have special ways of cooling themselves. Alligators open their mouths and let water evaporate from the moist surfaces within, and tortoises salivate over the front of the body.

Fertilisation is internal. Some species are viviparous, giving birth to live young, but most lay eggs. The egg has a **soft shell** and the embryo inside it is cushioned against damage (see box 6.1).

There is little parental care of the eggs and young, though crocodiles bury their eggs and, after hatching, protect the youngsters by taking them into their mouths. If you look at an adult crocodile you may see a young one staring at you from between its teeth.

Reptiles have left a remarkably full fossil record from which their evolutionary history has been reconstructed. They were the dominant fauna during the Mesozoic era and included the famous dinosaurs whose fossil remains can be seen in museums all over the world. Amongst the ancient reptiles were species that gave rise to the first birds and mammals, the two vertebrate classes which dominate most terrestrial environments today.

6.5   Birds (Class Aves)

Birds have been described as 'feathered dinosaurs' and they do indeed have many similarities with reptiles, for example scales on the hind limbs and many details of the skeleton.

The most obvious differences are the **wings** and **feathers**. The feathers, made of keratin and of intricate design, are assiduously preened by their owner who spreads oil over them from a gland near the anus. Feathers insulate and – because of the oil – waterproof the body. Their colours are important in camouflage, reproductive behaviour and aggression.

Another difference from reptiles is the **beak**. There are no teeth; food is swallowed whole. Seed-eating birds, such as pigeons, grind up the food in a special part of the stomach, the **gizzard**, which contains small stones. The

Box 6.1 Adaptations to life on land

Here we consider the problems that faced the first terrestrial vertebrates and the evolutionary changes that enabled them to survive on land.

Air is about a thousand times less dense than water, so if a fish is out of water the gills collapse and the filaments stick together. This reduces the surface area for gaseous exchange, so the fish dies. The **lungs** of terrestrial vertebrates always contain some air, which prevents them collapsing. Oxygen has to go into solution before it can enter the bloodstream, so the gaseous exchange surface in the lungs is kept permanently moist.

Air cannot support the body as water does, so a different method of locomotion is required. The fins are replaced by limbs (see figure 6.6). The limb of the terrestrial vertebrate – called the **pentadactyl limb** because it has five digits – is essentially the same in all terrestrial vertebrates from frogs to humans. However, it has become modified in some groups for particular purposes (look at the bird wing in figure 6.8 for an example). The vertebral column, spanning the gap between the fore limbs and hind limbs, is strong yet flexible and carries the weight of the body.

Terrestrial vertebrates are liable to lose water by evaporation from the body. This is prevented, or at least reduced, by having an **impermeable**

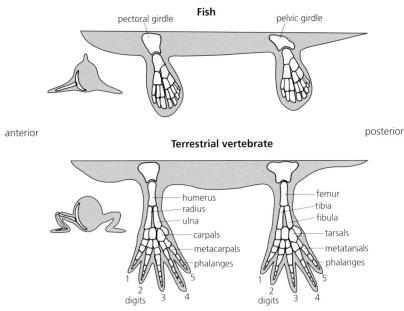

Figure 6.6 The paired fins of an ancestral aquatic fish (crossopterygian) and the limbs of an early terrestrial vertebrate to which they may have given rise in evolution. The transverse sections on the left show the splayed out stance typical of early forms and seen today in many amphibians and reptiles. (Redrawn from Goodrich E. S. 1930 *Studies on the structure and development of vertebrates*, Macmillan.)

skin and by reabsorbing water from the excretory waste. Ideally, the nitrogenous waste should be excreted in solid form (**uric acid**), as it is in reptiles and birds. The eyes are protected and kept moist by moveable upper and lower lids, and many land vertebrates – notably reptiles and birds – have a transparent third eyelid, the **nictitating membrane**, which closes over them like a shutter.

Water is not subject to the wide fluctuations in temperature that characterise terrestrial environments. The body temperature of fishes and most other aquatic animals is the same as that of their surroundings – that is, they are **ectothermic**. Many land animals are ectothermic too, but birds and mammals are **endothermic**, maintaining a body temperature independent of their surroundings.

An aquatic animal can shed its eggs and sperm into the surrounding water where fertilisation takes place (**external fertilisation**). This would be impossible for a terrestrial animal, unless it returns to water to breed, as amphibians do. In fully terrestrial animals, fertilisation occurs inside the female's body (**internal fertilisation**). The female may lay the fertilised eggs, which develop outside her body (**oviparity**), or she may keep them inside her body until the young are born (**viviparity**). Most reptiles and all birds are oviparous; most mammals are viviparous.

Eggs laid on land are susceptible to damage and desiccation. The eggs of reptiles and birds have a protective **shell** and, inside, a system of **extra-embryonic membranes** (so called because they are *outside* the embryo), which enable the embryo to feed, breathe and excrete until the egg hatches (see figure 6.7).

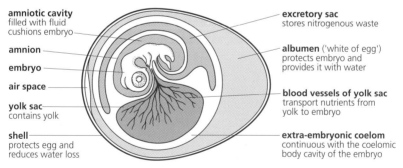

amniotic cavity
filled with fluid
cushions embryo

amnion

embryo

air space

yolk sac
contains yolk

shell
protects egg and
reduces water loss

excretory sac
stores nitrogenous waste

albumen ('white of egg')
protects embryo and
provides it with water

blood vessels of yolk sac
transport nutrients from
yolk to embryo

extra-embryonic coelom
continuous with the coelomic
body cavity of the embryo

Figure 6.7   The inside of a hen's egg with a developing embryo and extraembryonic membranes enclosing the amniotic cavity, excretory sac and yolk sac. The shell is permeable to gases. It reduces, but does not prevent altogether, evaporative water loss. As development proceeds, the amniotic cavity and excretory sac expand. The wall of the excretory sac, called the allantois, gets richly vascularised and becomes a gaseous exchange surface just under the shell. In eutherian mammals, it becomes the placenta. (Redrawn from Keeton W. T. 1980 *Biological science*, 3rd edition, Norton.)

gizzard is rather effective: a certain turkey is reputed to have ground up 12 steel needles in 36 hours!

Birds have many adaptations for flight (see figure 6.8). The most obvious adaptation are the wings with their long, overlapping **flight feathers** (see figure 6.8a). Flapping of the wings (**active flight**) is brought about by the contraction of powerful **flight muscles**, which run from the upper wing bone (humerus) to the breastbone (sternum) (see figure 6.8b). These muscles contain a lot of **myoglobin**, a pigment with a particularly high affinity for oxygen.

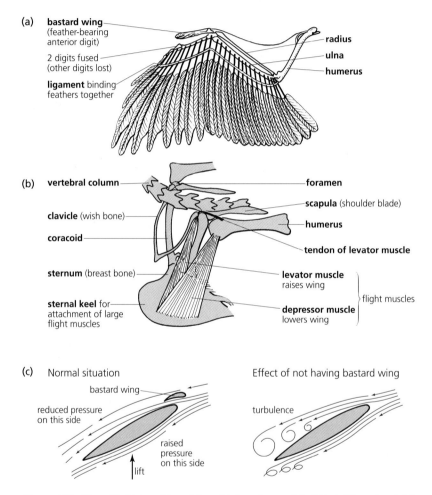

Figure 6.8 How flight is achieved in a bird such as a blackbird. The wing (*a*) is a modified pentadactyl limb with a reduced number of digits. The tendon of the levator muscle, shown in (*b*), runs through the foramen bounded by the scapula, coracoid and clavicle and is inserted on the upper side of the humerus. It raises the wing by working like a pulley. The sternal keel is an extension of the sternum for attachment of the large flight muscles. During gliding (*c*), the bastard wing and the feathers towards the wing tip smooth the flow of air over the top of the wing, preventing turbulence.

Once airborne, many birds glide (**passive flight**). The wings, held out from the body at an angle to the airflow, act as aerofoils holding the bird up. The bastard wings and the gaps (slots) between the feathers, particularly towards the tip of the wing, smooth the flow of air over the wing and prevent turbulence, which could cause the bird to stall (see figure 6.8c). Gliding is helped by having long wings with a large surface area. The wingspan of the wandering albatross can exceed 3.5 m. By making use of rising air currents, some birds can glide for well over 50 km without beating their wings at all.

Other adaptations for flight include the possession of **air sacs**, continuous with the lungs, which make the body more buoyant. The skeleton is light and the air sacs penetrate some of the bones. Many birds have prehensile (grasping) feet, enabling them to perch between flights or, in the case of predatory birds such as owls and eagles, to swoop and grasp their prey.

The lungs (see figure 6.9) consist of about a thousand narrow **parabronchial tubes** whose walls are riddled with minute **air capillaries** where gaseous exchange takes place. The parabronchial tubes are richly vascularised and the distance between the air and blood is very short, thus facilitating diffusion of gases. During breathing, air flows in and out of the air sacs and this helps to cool the body. The ventilation mechanism is complex, involving the movements of ribs and sternum.

The circulation is double, and the ventricle of the heart is completely divided into right and left sides by a septum, thus keeping deoxygenated and oxygenated blood entirely separate.

Birds are endothermic, generating metabolic heat energy, which is kept in the body by the insulating feathers. Water is conserved by reabsorbing it from the excretory matter and faeces, nitrogenous waste being released in the form of uric acid. The ability of birds to maintain a constant body temperature and conserve water helps to explain their wide distribution.

In many species, reproduction starts with the male acquiring and defending a piece of **territory**, followed by nest-building and **courtship**. Courtship bonds the sexes and synchronises the mating process. In courtship and territorial defence, **bird song** is important and the ability to vocalise as a means of communication must have been very important in the evolution of birds.

Fertilisation is internal and the eggs, similar to those of reptiles but with a hard shell, are laid in the nest and incubated by one or both parents. After hatching, the young are fed and protected by the parents until they can fly and fend for themselves.

Birds have excellent eyes and quick responses. A buzzard can detect the movement of a small mammal from a height of over 100 m and then swoop on it with considerable speed and accuracy. Behaviour, particularly during courtship, is mainly instinctive though learning plays some part, especially in the way young birds develop their songs.

Some birds stay in the same area all their lives, but many **migrate**.

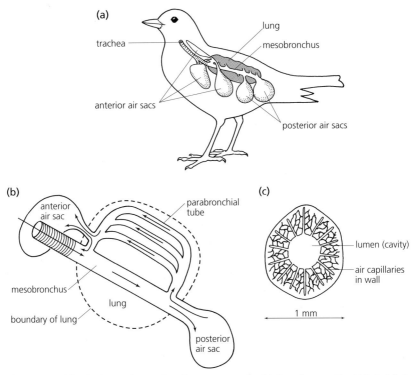

Figure 6.9 The lungs and associated structures of a bird such as a blackbird. (*a*) A general side view. (*b*) A diagrammatic side view in which the arrows indicate one of several routes by which air may flow through the lungs. (*c*) Transverse section of a parabronchial tube – gaseous exchange takes place in the walls, which are richly vascularised. (The general side view is redrawn from Hughes G. M. 1979 *The vertebrate lung*, 2nd edition, Carolina Biology Readers. The diagrammatic side view is redrawn from Schmidt-Nielsen K. 1990 *Animal physiology: adaptation and environment*, 4th edition, Cambridge University Press.)

Their ability to return to their original nesting sites after migrating is remarkable. The Arctic tern migrates from its summer breeding grounds in the Arctic to its feeding grounds in the Antarctic, a distance of 18 000 km. Migratory birds navigate by using the earth's magnetic field and the positions of the sun and stars. To do this, a human would need a clock, a compass and a map. A bird has all three inside its brain!

There are over 8000 living species of birds. Some have lost the ability to fly, and live on land or in water. The ostrich, for example, is a flightless bird with long powerful legs for walking and running. Penguins walk with an ungainly waddle but are superb swimmers, propelling themselves by their wings. Many birds are equally at home in air or on water, their webbed feet being used as paddles. Webbing of the feet is just one way in which the hind limbs of birds have become modified for different uses. Beaks, too, have become modified to cope with specialised feeding habits. These are good examples of adaptive radiation.

6.6   Mammals (Class Mammalia)

The Latin word *mamma* means breast or teat, and it is characteristic of mammals that the young are fed (suckled) on milk from **mammary glands,** modified sweat glands.

Mammals have **hair,** which is made of keratin and arises from pits (**hair follicles**) in the skin. Mammals are endothermic and in most species the hair insulates the body and helps to keep the body temperature constant, as does **subcutaneous fat** beneath the skin. Oil secreted by **sebaceous glands** in the skin keeps the hair greasy and helps to make the skin waterproof.

The teeth, unlike those of other vertebrates, are differentiated into specialised types: **incisors** for cutting, **canines** for piercing, and **premolars** and **molars** for crushing. The detailed structure of the teeth is related to the type of food that the animal eats. Thus carnivores, like lions and tigers, have large dagger-like canines for piercing and tearing flesh, whereas herbivores, like horses and deer, have molars with ridged surfaces for grinding plant food.

The limbs, instead of being splayed out, are in a vertical line under the body, making them more efficient at holding the body off the ground. Many mammals have long legs, enhancing their propulsive lever action. This, together with a flexible vertebral column, enables some mammals to move extremely swiftly. A cheetah can run, at least for short distances, at 100 km per hour.

In contrast to birds with their parabronchial tubes, the lungs are composed of numerous sac-like **alveoli.** The alveoli are enveloped by a dense network of blood capillaries, and it is here that gaseous exchange takes place. As in birds, the diffusion distance between the air and blood is very short. The lungs are ventilated by **ribs** and a **diaphragm.**

There is a double circulation, and the ventricle is completely divided into right and left halves for dealing with deoxygenated and oxygenated bloodstreams.

Reproduction and classification

Mammals are divided into three subclasses on the basis of how they reproduce:

- **Prototheria (monotremes)** are oviparous and lay eggs. The eggs hatch and the young are then supplied with milk by the mother. The only two living genera, the duck-billed platypus *Ornithorhynchus* and spiny anteater *Echidna*, are confined to Australasia.
- **Metatheria (marsupials)** are viviparous but the young are born at an early stage when they are unable to fend for themselves, so they are housed and suckled in a pouch (**marsupium**) on the lower part of the mother's abdomen. There is no true placenta, but the yolk sac

functions as one during the short gestation period. Marsupials are found only in Australasia and America.

- **Eutherians** (**placentals**) are viviparous and have a **true placenta** (formed from the allantois). The young are born at a relatively late stage when they are more capable of fending for themselves. Most mammals, including humans, belong to this group and they are found all over the world. There are about 4000 living species.

Eutherian mammals develop inside the mother's uterus, where the placenta provides an intimate association between the bloodstreams of the mother and embryo. Across it, the embryo obtains oxygen, nutrients and antibodies and gets rid of carbon dioxide and nitrogenous waste. The embryo is surrounded by a fluid-filled **amniotic cavity** (homologous with the one in the eggs of reptiles and birds), which cushions it.

Mammals, particularly eutherians, take great care of their young, protecting, feeding and training them. Behaviour, though partly instinctive, is readily modified by past experience – so it involves **learning**. Humans and their close primate relatives show a further type of behaviour, **insight**, which enables them to predict the outcome of their actions when faced with novel situations.

Mammals exploit many kinds of food and live in most kinds of environment (see figure 6.10). Their adaptive radiation is impressive. They can run, jump, burrow, dive, swim, fly, climb, swing and send faxes to each other. No other group of animals shows such diversity.

Figure 6.10 Eutherian mammals have successfully exploited the world's three main environments: land, water and air. This is illustrated here by a swinging white-handed gibbon (*Hylobates lar*), Atlantic spotted dolphins (*Stenella frontalis*) swimming, and a greater horseshoe bat (*Rhinolophus ferrum-equinum*) in full flight.

Maintaining biodiversity

So far, we have looked at biodiversity in terms of the variety of living organisms. Now we shall consider these organisms in their natural surroundings, but first we need to clarify a few ecological terms:

- **habitat** – the place where an organism lives
- **environment** – the surroundings of an organism including physical features and other organisms
- **population** – a group of organisms of one species living in the same area so that mature individuals have the potential to reproduce with others in the group
- **community** – organisms from many species which interact together within a common environment
- **ecosystem** – a community of organisms together with the physical environment where they live
- **biome** – collective name for communities with the same appearance living in similar environments but in different parts of the world.

Tremendous diversities of life forms occur in a wide variety of ecosystems. We need to understand the threats to these ecosystems and how biodiversity may change in the future.

7.1   Global patterns of diversity

How many species?

After centuries of biological investigation, we are still discovering new species and even new kingdoms. Two of the last great places of mystery are the tropical forest canopy and the ocean abyss. The deep oceans are so vast that we have little idea what is down there. The few special automatic cameras which have been sent to the ocean floors have glimpsed many strange animals, most of them never seen before; some look like extras from a horror movie.

So how many species are there in the world and how many remain to be discovered? We only have a rough idea (see table 7.1). Most terrestrial vertebrates and non-tropical plants have probably been discovered.

Table 7.1   The approximate number of species known, and the estimated total, for several important groups.

Taxon	Number of species (thousands)	
	Known	Estimated
Bacteria	4	1000
Protists	42	200
Algae	40	300
Fungi	77	1000
Viruses	5	400
Bryophytes	16	26
Pteridophytes	10	20
Seed plants	240	300
Nematodes	20	400
Molluscs	72	200
Crustaceans	43	150
Arachnids	80	750
Insects	980	8000
Vertebrates	47	50

Source: (Based on data from *World Conservation Monitoring Centre Report on Global Diversity* 1992 edited by B. Groombridge, and *The United Nations Environment Programme Global Diversity Assessment* 1995 edited by V. H. Heywood.)

However, our knowledge of tropical species is still patchy and of deep sea invertebrates only just beginning. We can only make wild guesses as to how many bacteria are still undiscovered.

One way to estimate how many bacterial species are present in a soil or other sample is a DNA pairing test. The DNA extracted from the bacteria is heated until all the double strands separate. It is then cooled and the time for all the strands to re-pair is noted. If all the DNA strands are similar, single strands quickly find a complementary strand as the DNA re-forms double helices. If the DNA is varied, strands take longer to re-pair. The re-pairing time reflects how many different species are present. A single gram of soil tested this way indicated the presence of 10 000 different types of bacteria! That sort of abundance raises all sorts of questions about bacterial distribution. Are they so evenly distributed that every area tested in a region would have the same 10 000 species, or could a gram of soil only a few metres away have a *different* 10 000 species? This test does not identify particular species, it only indicates diversity. From such experiments, and from table 7.1, it is easy to see why microbiologists who study prokaryotes think that their organisms are so worthy of study.

Biomes

If you took a trip round the world, you would find both familiar and unfamiliar types of wildlife wherever you went. You would see many different types of vegetation: evergreen coniferous forests, grasslands, *Sphagnum* bogs and deserts with little visible vegetation. You would recognise *similar* communities such as coniferous forests or tidal rock pools in different parts of the world, and these make up biomes. However, these similar communities would contain *different* species.

The size and form of the plants in terrestrial biomes is very distinctive, so terrestrial biomes are classified on their vegetation type. For example, tropical forest, coastal mangrove swamp, desert and dry temperate grassland (steppe and prairie) are all biomes. For aquatic ecosystems the descriptions are more varied, but usually include some aspect of the physical environment such as stream, deep lake, ephemeral pool, continental shelf, barrier reef and so on.

In any biome, the source of food for the heterotrophs is important. Most terrestrial and aquatic communities rely on photosynthetic primary producers. A few communities rely on prokaryote chemoautotrophs such as those in deep ocean vents, sulphur springs and subterranean caves with a hydrogen sulphide source (only one such cave system has so far been found). Some ecosystems rely entirely on external food sources: spiders on barren volcanic hillsides live on insects blown in on the wind; animals in deep oceans feed on organic matter sinking down from the plankton above.

Terrestrial biomes

The vegetation of a terrestrial biome is determined by the climate, especially temperature and rainfall. Average temperature decreases from equator to pole and seasonal extremes increase. It is an oversimplification to say there are no seasons in the tropics: rainfall is often higher at certain times of the year. However, seasonality is slight compared to the summer–winter cycles of temperate and boreal zones and the months of continuous darkness or daylight in polar regions.

Because climate changes with latitude, the biomes are arranged in latitudinal patterns. Rainfall does not follow the same zonal pattern as temperature; it varies depending on prevailing winds, and the positions of oceans and mountain ridges. Rainfall therefore adds complexity to biome distribution, disrupting what would otherwise be a clear latitudinal zonation. The major terrestrial biomes can be plotted on a chart of increasing seasonality and decreasing rainfall (see figure 7.1). Other biomes such as elfin wood and alpine meadow are associated with mountains.

Hot, wet biomes have the highest diversity of species, and cold, dry ones the lowest. This applies to most phyla in the plant and animal kingdoms and underpins the general observation that species numbers

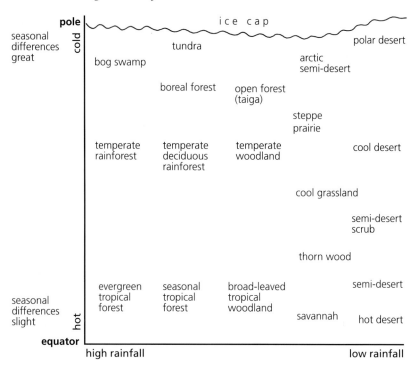

Figure 7.1 Distribution of the main terrestrial biomes on climate gradients. Seasonality increases and temperature decreases from equator to pole, while variation in rainfall occurs at all latitudes.

decline from equator to pole. We do not know enough to say if this trend also holds for bacteria and other cryptic microorganisms.

Tropical forest diversity

Tropical forests now cover about 7% of the land surface (half their natural range), yet they probably contain between 50% and 90% of all species. Huge trees form the canopy with smaller trees and bushes beneath. Plant diversity is great both in plant form (trees, shrubs, climbers and epiphytes) and in species. A few hectares of forest can have 700 species of tree, each with associated insects, birds, bacteria and so on.

The reason for this huge diversity is not clear. One view is that tropical forest is ancient and provides a stable environment where species evolved into specialists with very precise roles (**niches**). The smaller the niches, the more species can 'fit' into the ecosystem and the diversity of plants provides many niches for other organisms.

A different view is that tropical forest is neither ancient nor stable. Until a few million years ago, the tropics were hot and dry and did not maintain lush vegetation. We now think that the series of ice ages which have racked the globe over the last 2 million years affected the forest more than was first supposed. About every 120 000 years, the glacial advance dried out the lower latitudes, and savannah grassland spread across the tropical lowlands, dividing the tropical forest into small fragments called **refugia** (local refuges for species). These would last some 100 000 years, during which time their isolation would lead to allopatric speciation (see page 3), with different species evolving in each refugium. During the shorter, warmer and wetter interglacials, the forests advanced again and recolonised the grasslands producing an increase in diversity as the forests merged.

Diversity in tropical species also results from coevolution between insects and angiosperms (see box 5.2 on page 80). When angiosperms diversified during the Cretaceous, 100 million years ago, they interacted with animals which transported pollen from flower to flower and dispersed seeds and fruits. Insects carry pollen to flowers of other plants in the species, so each insect-pollinated tree species can survive at low density scattered among many other species. The angiosperms provide a range of microhabitats for insects as pollinators, leaf and fruit eaters, bark dwellers and so on.

Tropical diversity is under threat as forest is being lost – cut down for timber or paper making and burned for farming land. Recently, this has been happening at an alarming rate in most parts of the tropics.

Tropical coral reef

Coral reefs are the tropical forests of the sea. They are probably the most diverse of the marine biomes (although the abyss may surprise us yet). Individual reefs are built up over thousands of years from the calcium carbonate skeletons of colonial cnidarians (see page 56). Around this structure live a tremendously colourful mix of invertebrates and fishes. The invertebrates include echinoderms, bivalve molluscs, shrimps and cephalopods.

Corals use their tentacles to catch particles which drift down from the phytoplankton above. Some cnidarians have thousands of photosynthetic dinoflagellates living endosymbiotically in their tentacles, giving them a greenish appearance. Because these protists require light, corals live in clear waters down to at most 50 m depth. Turbidity destroys the symbiosis by reducing the protists' rate of photosynthesis.

Muddy waters have a further damaging effect as they clog up the reef and smother it to death. Many areas of the Great Barrier Reef off the coast of Australia have been destroyed by sediments from farmland that have been washed into the rivers and then out to sea. Oil spills too do serious and lasting damage to corals and thus the whole reef. Another recent threat is a large multi-armed starfish called the crown of thorns, which has become very abundant and eats the corals.

7.2 Extinction

The threat to biodiversity

The number of species in an area is always changing on several different timescales. Species numbers can increase rapidly due to **immigration** of individuals of new species from an adjacent area, **introduction** of an alien species from a distant locality (due to human influence) or more slowly by **evolution** of a new species. Species are lost from an area due to **emigration** of individuals of a mobile species or **local extinction** – the death of the entire population in an area.

Extinction of a local population may have little consequence for biodiversity if the population is a common, widespread species, but it may result in reduction of genetic diversity or loss of a subspecies in a rare or scattered taxon. If the population was the last on the earth, then its loss means global extinction. Extinction is certainly for ever, but it is not a rare event in the earth's history. For every species alive today, there are at least a thousand extinct species.

Mass extinctions

The fossil record has only preserved about 0.005% of multicellular species. It is extremely biased: hard-shelled invertebrates and vascular plants are well represented, insects and soft-bodied invertebrates poorly so. Some unicellular eukaryotes – for example dinoflagellates, diatoms and foraminifera – are well recorded, but most protists and prokaryotes have left no trace. Despite the inadequacies of the fossil record, it is possible to build up a picture of past species diversity as having a background level of extinctions plus several occasions when most species disappeared (**mass extinctions**).

The worst mass extinction was at the end of the Permian, 245 million years ago, when 54% of families, 80% of genera and 95% of species of marine organism vanished. The most famous mass extinction occurred at the Cretaceous–Tertiary boundary when the last dinosaurs became extinct. Only about 15% of marine families disappeared, but 75% of plant species and 40% of land vertebrate families were lost.

Causes of mass extinction

At least five major mass extinctions are recorded in the fossil record. The Cretaceous–Tertiary boundary event differed from the others in that mainly terrestrial species, rather than marine ones, became extinct. Many geologists now accept that a huge catastrophe in the form of a meteor impact threw the global environment into chaos. The meteor was a large one and may have crashed near present-day Mexico. The debris thrown into the atmosphere would have blocked out the sunlight, causing a severe drop in temperature as well as depriving photosynthetic organisms of their energy source. Without primary producers, whole food chains would have collapsed.

The Permian mass extinction took about 6 million years, a much slower process than a meteor impact. During the Permian, the huge continental plates of the earth's crust joined into a supercontinent, changing or destroying continental shelves where many marine organisms lived. Greatly increased volcanic activity was associated with the plate movement. At the same time, changes in sea currents and atmospheric movement caused climate changes. These accumulated effects led to the extinction of many species.

Less extreme extinction events may have been triggered by oscillating phases of glaciation. For the last 2 million years, the earth has been glacial with long cold phases of about 110 000 years interspersed with shorter warm spells of about 12 000 years. There have been over 15 glacial–interglacial cycles so far. In higher latitudes, advancing ice sheets destroyed most of the vegetation down to southern Britain and across North America and Europe. This devastating effect on diversity, 'wiping the slate clean' over and over again, has depleted the flora of Britain and Ireland. Species migrated south across mainland Europe in advance of the glaciers. When the ice retreated, some species never made it back to Britain before the English Channel separated it from the rest of Europe.

The current warmer interglacial started about 10 000 years ago. We could be due for another glacial in two to four thousand years' time. That is unless the greenhouse effect from high levels of atmospheric carbon dioxide and other pollutants does not trigger a new climate pattern first.

The diversity gradient from equator to pole may have been steepened by the glacial cycles if the instability at high latitudes resulted in diversity loss while the more stable, but still disturbed, pattern in the tropics produced a relatively high species diversity. Because some mass extinctions seem to have been caused by glaciation, our present sequence of glacial cycles raises disturbing questions. Do we inhabit a planet suffering recent (i.e. last 2 million years) upheaval, with reduced biodiversity, especially in higher latitudes? Are we further reducing biodiversity by our activities to a dangerously low level?

Life on earth seems to recover rather well from mass extinction events. Surviving species undergo rapid radiative evolution and new species form. Of course, groups which become extinct never re-evolve (see box 7.1), but other varied forms arise to take their place.

If we cause a mass extinction in the next few hundred years, can the world recover? The geologists' answer is 'almost certainly yes'; biologists say 'yes, but look what we will have lost'. The problem is that recovery will take millions of years and the earth might not be a very congenial place in the meantime – indeed it might not be able to support human life. We might become extinct too.

Box 7.1 Extinct organisms

Many taxa, even whole phyla, have gone extinct during the history of life on earth. Here are four examples.

1 Trilobites (Phylum Trilobita) were maritime arthropods; at least some were carnivores. Closest living relatives include shrimps and horseshoe crabs (see page 00)

2 Giant clubmosses (Phylum Lepidodendrales) were swamp forest trees. Their crushed remains fossilised to form coal seams. They became extinct at the end of the Carboniferous about 290 million years ago. Their closest living relatives today are tiny clubmosses (Lycopodium) and quillworts (Isoetes).

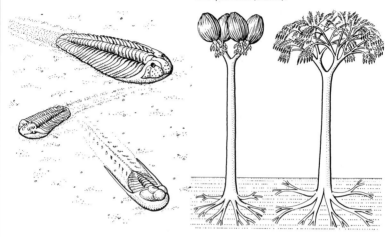

3 Bennettites (Phylum Cycadeoidales) were gymnosperms with stumpy trunks and tough pinnate leaves. They had mysterious cones which may have been pollinated by insects and possibly smelled of rotting dinosaur flesh to attract carrion beetles! They lived in the Cretaceous and nothing like them survives today.

4 Oviraptorids, the eggstealer dinosaurs (Order Saurischia) were named because they were sometimes found near eggs.It is now known that they guarded their own nests rather than destroyed others. They lived in Mongolia in the late Cretaceous and may have died out because of the meteor impact (see page 00). Nearest living relatives today are birds.

0.5 m

Trilobites adapted from Winson E. 1978 in McKerrow W. S. (ed.) *The ecology of fossils*, Duckworth & Co. Ltd; giant clubmosses Hirmer M. 1927 *Handbuch der Paläobotanik*, R. Oldenbourg; cycadeoid from Watson J. 1991 in Watson J. & Sincock C. A. 1992 *Bennettitales of the English Wealden*, monograph of the Palaeontographical Society; and oviraptorid from Lambert D. 1989 *Dinosaur data book*, Facts on File/British Museum (Natural History).

Extinction rates now

Are we in the middle of another mass extinction event? Many conservationists would say we are. Species do seem to be vanishing at an alarming rate,
but what was the rate in the past and can we compare it with the present?

The average species survival time in the fossil record is 4 million years.
Thus, on average, 25% of species go extinct every million years. If we take
the number of species today as 5–30 million, then the usual extinction rate
should be between one and seven species per year. During periods of mass
extinction, that rate increases three to four fold: 75% to 95% of species go
extinct, so mass extinction rates are perhaps 15 to 30 species per year.

There are numerous records of extinctions in the last few centuries.
Many were animals actively hunted during that time: the dodo of Mauritius,
the passenger pigeon of America and the thylacine in Tasmania to name but
three. Known extinctions are only the tip of the iceberg: many species will
have vanished before they have been discovered.

We can estimate current extinction rates from the damage and destruction done to habitats. There is a species–area relationship which is used to
calculate species loss. For example, if 90% of an area of forest is destroyed,
then 50% of the species there go extinct. Estimates of current extinction rates
vary widely depending on how diverse tropical forest is assumed to be. The
tropics, where forest destruction is high, contain the most diverse biomes, so
most species loss occurs there. Some botanists estimate that 2000 plant
species a year are disappearing from these forests. Estimates of total global
species loss range from 4000 to 300 000 per year.

Some areas of tropical forest are much richer than others. If those areas
could be identified and preserved, then much diversity could be saved even
if the rest of the forest was damaged.

Causes of recent extinctions

It is impossible to say what caused extinctions in the past, but we have a
better idea about recent losses, from historical records and archaeological
investigations. Figure 7.2 charts these causes, and you can see there is a
human influence in many of them.

The decline to extinction may be abrupt when the final population is
destroyed by a catastrophe, or it may be slow if the population is still
breeding but mortality rates are just slightly higher than survival rates.
Sometimes it is harder to detect the causes in a slow decline of a species.
Extinctions probably result from a number of causes which add up to just
too much pressure on the species.

Island species are especially at risk because they have small populations contained in a small area. Isolated islands have special ecologies with
an unusual mixture of organisms. Many islands have a number of **endemic**

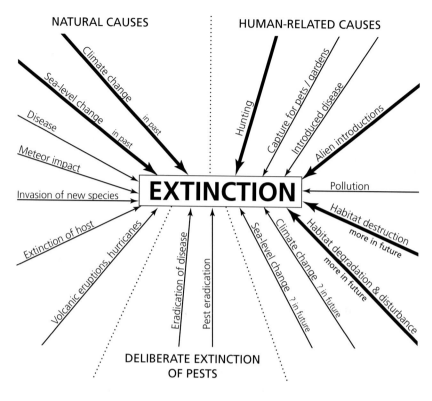

NATURAL CAUSES

HUMAN-RELATED CAUSES

EXTINCTION

**DELIBERATE EXTINCTION
OF PESTS**

Natural causes
Climate change, in past
Sea-level change, in past
Disease: sweet chestnut (USA)
Meteor impact: last dinosaurs,
 rare but with extensive consequences
Invasion of new species, competition
Extinction of host: dinosaur gut parasites (?)
Volcanic eruptions, hurricanes
 etc., especially on islands

Human-related causes
Hunting: passenger pigeon, moa,
 rhinos (nearly), tiger (nearly)
Capture for pets/gardens: tortoises,
 parrots and macaws, orchids
Introduced disease: New Zealand quail
Alien introductions, especially on islands:
 40spp. including *Partula* tree snail
 (killed by another snail in Polynesia) and
 cichlid fish in Lake Victoria (killed by Nile Perch)
Pollution
Habitat destruction, more in future:
 Grand Cayman thrush, Ryuku wood pigeon (Japan)
Habitat degradation and disturbance, more in future:
 stumptooth minnow (Mexico), Israeli painted frog
Climate change
Sea-level change

Deliberate extinction of pests
Disease eradication: smallpox virus
Pest eradication: Falkland Island wolf

Figure 7.2   The causes of extinction of species with some named examples where it is known to be the main cause. The thickness of the arrows indicates the relative importance of the phenomenon as a threat to species.

species, which evolved on the island and are found nowhere else. These endemics are particularly vulnerable because they are easily affected by natural disasters, such as hurricanes or volcanic eruptions (many isolated islands are volcanic). Human influence from hunting or the introduction of alien species also pose threats. Many island species have been lost because of the activities of cats, goats, pigs or dogs. In fact, about 75% of recorded extinctions have been of island species. In the future, extinctions on continents will increase as more and more habitats are destroyed.

Sometimes an abundant species can be wiped out. In the 1700s, there were 5000 million passenger pigeons in North America. The last one, called Martha, died in a zoo on 1 September 1914. Although habitat destruction must have been responsible for the early decline in passenger pigeon numbers, hunting also played its part.

In the North American broad-leaved forests, sweet chestnuts suddenly started to die in the nineteenth century as the result of infection by a virulent pathogenic fungus. There were almost as many chestnuts as passenger pigeons, but in the last hundred years all have gone except for a few stumps. These sprout green shoots, but they never survive more than a year or two before the fungus attacks them again. Soon, with no new seeds to start the next generation, all the trees will be dead.

Ecosystems contain complex associations of organisms and some are more important in maintaining the ecosystem than others. The important ones are called **keystone species**. If a keystone species goes extinct, then others may follow. In the Florida Everglades, crocodiles are a keystone species. Many other animals rely on these large reptiles to maintain deep water. In the Antarctic oceans, krill (tiny arthropods) are an important link in the food chain for many larger species including whales. Krill is a keystone group over which there is some concern because fishing fleets are starting to exploit them, leaving less krill for other sea creatures. In the northern hemisphere, the same is true of sand eels.

Over the last 50 000 years or so, the main human threat to biodiversity has been from hunting and from the introduction of alien species. Now, habitat destruction is becoming the major threat, with the potential to destroy many species at once. This, more than anything, suggests we need a global policy to deal with threats to biodiversity. One problem is that the most natural biomes usually occur in underdeveloped countries which want to increase their income. Many industrial countries have already destroyed most of their ecosystems and are not very good at protecting what is left. The pressure from large multinational companies to allow mining or timber-felling is strong. Sadly, money is a powerful short-term goal for many people.

Box 7.2 Gaia

In the 1970s, James Lovelock suggested a new way to look at the earth. The **Gaia hypothesis** asserts that there is a close 'working relationship' between the living organisms and the physical environment of the earth. (Gaia is the Ancient Greek title for Mother Earth.) The whole system works like a self-adjusting organism which results in the continuation of favourable conditions for life. Just as an individual (a single mammal, for example) breathes to maintain a regular input of oxygen and output of carbon dioxide, feeds to obtain organic material and energy, and regulates its body temperature, so the organic and chemical aspects of Gaia function to maintain a favourable state for life.

Gaia can be thought of as a buffered superorganism, but she is not unchanging. There have certainly been important developments on earth in the past. For example, the early atmosphere probably had no oxygen in it, but once prokaryotes began to split water molecules during autotrophy, oxygen began to build up in the atmosphere. At first, Gaia compensated by combining the oxygen with iron. As a result, huge red iron oxide beds are found interspersed with early cyanobacteria layers. However, the oxygen levels did rise in the atmosphere and earth moved on to a different phase in its evolution.

Lovelock points out that the very abundance of life on earth is fundamental to the way the earth system self-regulates. Threats to this abundance due to increased human activity (including pollution, atmospheric changes to carbon dioxide and destruction of ozone) could trigger a threshold effect which pushes Gaia into an entirely different equilibrium. Thus biodiversity increases the stability and resilience of Gaia.

7.3   Conservation

Does species loss matter?

Science fiction novels suggest some interesting scenarios for the earth. Some describe situations where most humans on earth are gone – usually in a nuclear war, sometimes after an ecological disaster. In others, human populations triumph but there is no wildlife left – people live on yeasts grown in hydroponic tanks. Neither alternative is particularly attractive. So is there a third option in which humans can live in the equilibrium of Gaia (see box 7.2)?

There are many reasons why we should maintain the earth's biodiversity:

- It gives us valuable medical products – a number of important drugs occur in plants (aspirin, for example) and many peoples use herbal medicines.
- We depend on biodiversity for other useful materials – we already obtain cloth fibres, timber, resins, glues, rubber and dyes from plants and animals.
- Many wild species are related to our crop plants and their genes could be used to improve disease resistance, flavour and vigour in our crops via cross-breeding or genetic engineering.
- Forests maintain the atmosphere, producing large amounts of water vapour and oxygen and taking in carbon dioxide. Without forests, climate change would increase.
- Loss of biodiversity decreases global stability – mass species loss would result in imbalance in ecosystems and may lessen human survival.
- We enjoy wildlife – ecotourism, rambling and bird-watching are all popular.
- The aesthetic argument – the living world is beautiful and it is our privilege to enjoy it. The beauty of a snow leopard or an orchid are obvious to most people, but *Paramecium* is also beautiful and so is a woodlouse.
- The ethical argument – all life has intrinsic value, even harmful organisms such as mosquitoes which carry malaria.
- We have a responsibility to our descendants to pass on a world worth living in.
- We are earth's stewards and have a moral responsibility to care for our fellow creatures. Or, to put it in religious terms, we have a responsibility to God to maintain the Creation.

Conserving biodiversity

The plight of rare mammals attracts much publicity. Saving pandas, tigers, elephants and whales all raise concern and create good TV programmes. Much is often expected of captive breeding in zoos for keeping these species from extinction. But zoos are not the long-term answer. They cannot maintain knowledge in a herd of elephants of how to survive the dry season; they cannot teach a monkey how to live in the forest, or a crane how to avoid being eaten by a bobcat. Many animals, when re-introduced into the wild, have little idea how to survive the dangers and complexities of a natural environment. Zoos are best for the short-term rescue of endangered species while their habitat is being salvaged. Invertebrates survive better as they have less complex behaviour patterns to lose while in captivity. There is little point in keeping species in zoos if they can never be released because their natural habitat has gone.

Large primary carnivores and big grazers should be conserved in their own habitat. **Preservation** of their ecosystems will result in hundreds or thousands of other species being protected too. Unfortunately, habitat preservation often conflicts with human interests. Land is used for farming, housing, roads, timber production, hydro-electric dams and so on. Even in Britain, where most of our natural habitats have already gone, 5% of Sites of Special Scientific Interest (SSSIs) are destroyed or damaged every year.

To preserve biodiversity, local people must be convinced of its worth and involved in the process. Enforced rules for conservation are seldom successful. Sometimes local people are eager to maintain the ecosystem, but government and business have other ideas. The native Amazonian Indians and rubber tappers, for example, have argued long and hard for their forests. They know that felling tropical forest is only a short-term economic gain; that the harvesting of nuts, fruits and rubber raises more than a one-off clearance for timber. In temperate forests too, wild mushroom harvesting can be more economically viable than felling for timber.

The African elephant

The fate of the African elephant, *Loxodonta africana*, illustrates well the fluctuating fortunes of a species in conflict with humans.

The elephant is the largest surviving land mammal (see figure 7.3). It has a strong matriarchal society with long-lasting family groups consisting of a mature female, her sisters and daughters and their offspring. The males, once they reach the age of ten years or so, leave the group to join other bachelor males. Large, sexually mature bull elephants move among family groups when a female becomes fertile and ready to mate.

Figure 7.3 A small family group of elephants in Amboseli, Kenya.

Young elephants remain for many years with the matriarch (all their lives if they are females) learning where the best feeding sites, water, salt sources and mud wallows are throughout the year. Behaviour changes during severe droughts, which may be several years apart; the matriarch leads the elephants to the best water sources where they dig for moisture.

Elephants are a keystone species: they create and enlarge watering holes by trampling the mud and carrying it away after wallowing. Their paths to waterholes also channel the rains. Many animals rely on these watering holes. Elephant dung provides habitats for dung-beetles, replenishes the fertility of the soil and distributes tree seeds. Elephants also open up scrubland, providing grass for grazing herds.

Elephant behaviour is learned in the herd. This knowledge could not be maintained in a zoo. After generations of captivity, elephants would have problems if returned to the wild. The loss of matriarchal guidance results in delinquent teenage elephants showing unusual behaviour.

Elephant populations have fluctuated considerably during the last century. Early in the 1900s, elephants were prized trophies for big-game hunters. They have always been hunted as a food source (even the woolly mammoths suffered this fate). As poaching was prevented and game hunting declined, populations boomed and by the 1960s scrublands were changing to open savannah and elephants were entering farmland for food.

Some conservationists argued that elephants were destroying their habitats by overgrazing, and elephants were culled in several countries. Whole matriarchal units were shot from helicopters – a terrible and distressing occurrence as elephants are intelligent and seemed to know what was happening. When elephants find bones of neighbours and relatives in their wanderings, they stop and caress them with their trunks. Their understanding of death is a mystery to us, but they are not unaffected by it.

In the early 1970s, several years of drought hit Central Africa. Many elephants died especially in countries where herds were not culled. The weaker, elderly and young animals died – not whole family groups as during culling. Family continuity was maintained; about 80% of the population, including the strongest and the best leaders, survived.

The drought ended just as the international price of ivory (elephant tusks) soared. Suddenly, poaching of elephants for profit and money to buy guns became widespread, especially in Central Africa, where guerrilla warfare was common. After 1973, elephant numbers plummeted to only 10–15% of past numbers. Early targets were bulls and matriarchs, but eventually even youngsters with small tusks were killed. The African elephant was threatened with extinction by the year 2000.

Something had to be done quickly since protecting elephants from organised groups of poachers was expensive and dangerous – several rangers lost their lives. In 1989, the Convention on International Trade in Endangered Species of Wild Fauna and Flora (CITES for short) banned the sale of ivory. Ivory prices collapsed and poaching declined. Not every

African country agrees with this ban, especially those in the south where poaching was less severe and elephant numbers remained higher. Many wildlife reserves obtained money for their elephant tusks and were reluctant to lose this income. Although some countries (like Kenya) burn the ivory they have confiscated, others are stockpiling it in case ivory trading resumes. They argue that labelling each tusk will prevent illegal trade and poaching; others fear elephant poaching will increase out of control again.

This story demonstrates the conflicts which may arise over management of wildlife:

- **Ecotourism** provides income for countries with safari parks and lots of large mammals. Management of wildlife for ecotourism often loses goodwill when local people feel they have lost control of their land.
- People want to kill animals and gather plants for food as they have always done. When they are allowed to profit from wildlife they care for it as a **renewable resource** (a resource which, if carefully managed, continues to provide without depleting the stocks).
- Elephants are dangerous (especially males) and damage crops – they sometimes kill people when invading farmland. Deaths in Kenya alone reached 30–40 per year in the 1980s until a new policy of dealing with problem animals reduced this to about 15 per year.
- Elephants require space and so do humans, and populations of both are rising. In Zimbabwe, for instance, elephant numbers rose from 4000 in 1900 to 76 000 in 1990; in the same time human populations rose from 0.5 million to 11 million.
- A few top wildlife managers and a lot of poachers want to take the ivory from parks for profit. Resumption of the ivory trade would encourage poaching. Deterring poachers is expensive, so ivory profits from legal culling in the protected parks would probably end up being spent on saving the elephant from being endangered again.

There will always be conflict between land for wildlife and land for farming. If people are to protect their wildlife, then they must be allowed to benefit directly from saving it. This applies in the Amazonian tropics or the Russian boreal forest just as much as in the African savannah.

7.4 What of the future?

The future will be challenging. In the next 50 years, we will know how increases in atmospheric carbon dioxide alter the climate. If global warming occurs, species excluded from cold biomes may survive at higher latitudes. The tropical zone will expand and the boreal and Arctic zones will be squeezed. In past glacial cycles, we know that species migrated long distances as the climate changed. Migration of plants in the future will be difficult as few continuous corridors of land are available. Many areas are covered in farmland where invading plant species will not find room to grow. A new challenge will be how to save species which inhabit areas undergoing climate change. **Habitat management** must maintain ecosystems in unfavourable, new climatic conditions or move whole ecosystems to new areas where the climate is more suitable.

As the human population increases and resources dwindle, there will be more pressure on remaining ecosystems. Governments will come under increasing pressure to exploit land for industry, housing, food and jobs.

In 50 years we will know if the tropical forest can be saved. We will also have a more complete record of recent global extinction rates that will show whether we are winning the battle to conserve species and ecosystems.

Further reading

Alexander R. McN. 1990 *Animals*, Cambridge University Press.

Attenborough D. 1979 *Life on earth: a natural history*, William Collins Sons & Co. and BBC Books.

Attenborough D. 1995 *The private life of plants*, BBC Books.

Buchsbaum R. *et al.* 1987 *Animals without backbones*, University of Chicago Press. (Also available as a Penguin.)

Cadogan A. & Hanks J. 1995 *Microbiology and biotechnology*, Thomas Nelson & Sons.

Dawkins R. 1976 *The selfish gene*, Oxford University Press.

Kershaw D. R. 1988 *Animal diversity*, Chapman and Hall.

Lewington A. 1990 *Plants for people*, Natural History Museum Publications.

Lowrie P. & Wells S. 1994 *Microbiology and biotechnology*, UCLES Cambridge Modular Sciences and Cambridge University Press.

Marshall P. T. & Hughes G. M. 1965 *The physiology of mammals and other vertebrates*, Cambridge University Press.

Mitchell A. W. 1986 *The enchanted canopy: secrets from the rainforest roof*, William Collins/Fontana.

Myers N. (ed.) 1994 *The Gaia atlas of planet management*, Gaia Books.

Ramsay J. A. 1968 *Physiological approach to the lower animals*, Cambridge University Press.

Reiss M. J. & Chapman J. L. 1994 *Ecology and conservation*, UCLES Cambridge Modular Sciences and Cambridge University Press.

Romer A. S. 1959 *The vertebrate story*, University of Chicago Press, 1959. (Also available as a two-volume Penguin under the title of 'Man and the Vertebrates'.)

Schmidt-Neilsen K. 1972 *How animals work*, Cambridge University Press.

Spellerberg I. F. & Hardes S. R. 1992 *Biological conservation*, Cambridge University Press.

Young J. Z. 1981 *The life of vertebrates*, Oxford University Press.

Index